Asteroids

Asteroids

How Love, Fear, and
Greed Will Determine
Our Future in Space

Martin Elvis

Yale UNIVERSITY PRESS
New Haven & London

Published with assistance from the Louis Stern Memorial Fund.

Yale University Press books may be purchased in quantity for educa-
tional, business, or promotional use. For information, please e-mail
sales.press@yale.edu (U.S. office) or sales@yaleup.co.uk (U.K. office).

Set in Galliard Old Style type by IDS Infotech Ltd., Chandigarh, India.
Printed in the United States of America.

Library of Congress Control Number: 2020948370
ISBN 978-0-300-23192-2 (hardcover : alk. paper)

A catalogue record for this book is available from the British Library.

This paper meets the requirements of ANSI/NISO Z39.48-1992
(Permanence of Paper).

10 9 8 7 6 5 4 3 2 1

Contents

Part IV. Opportunity

Preface

In the past few years the idea of mining the asteroids for their resources has been much in the news. Breathless stories described billionaires and movie directors investing in start-ups that would soon generate fortunes from precious metals gleaned from the asteroids. Several notable personalities declared that the first trillionaires would be asteroid miners. As a scientist I found these claims dubious. I wanted them to be true, but I had to investigate. This book is the result.

By starting from the basic motivations that make asteroids appealing to those who would exploit them—love, fear, and greed—my hope is to give the reader a fuller perspective of the whole issue. There is more to the story than just the technical challenges about how to identify the truly valuable asteroids and how to proceed with mining these; there are also important questions about the markets for asteroid resources, the economics involved, the legal framework that will be adopted, and the input that will be needed from policy experts and diplomats. There are even ethical issues to be considered. So many skills will be necessary to make asteroid mining a reality that no one can be a true expert in all aspects of the endeavor, but the aim of this book is to provide an introduction to them all.

After setting the scene with an introduction to the science of asteroids, I'll proceed to our motives for going to them: love of finding things out, fear of being destroyed by a killer asteroid, and the greed for profit. The chapters that follow will explain the means by which we will go about mining the asteroids, leading to the final section, in which we will examine the opportunities they could yield. Finally, we'll look at what we can expect in the long run if asteroid mining takes off.

Acknowledgments

No one can be a true expert on every aspect of asteroid mining. I have relied heavily on the knowledge and skills of many colleagues in writing this book. Deep thanks and gratitude are due to all those who gave their time to answer my queries. At the Center for Astrophysics | Harvard & Smithsonian, I thank my colleagues Jonathan McDowell, JL Galache, Peter Vereš, Kim McLeod (on sabbatical from Wellesley), Maria McEachern, and Charles Alcock, as well as my students Charlie Beeson, Thomas Esty, Chris Desira, Nina Hooper, Sukrit Ranjan, Anthony Taylor, and Matt Ontiveros. I also thank BC Crandall for early encouragement; Alanna Krolikowski and Tony Milligan (KCL), my collaborators on policy and ethics; Matt Weinzierl, Martin Stürmer, Ian Lange, Angela Acocela, and Anette Mikes of the Harvard Business School for advice in business and economics; John S. Lewis, the founding father of asteroid mining, for advice on resources; Jessica Snyder, Mark Sonter, Benjamin Lehner, and many others for their deep knowledge of asteroids; Rick Binzel, Francesca DeMeo, for the naming of 9283 Martinelvis, and Bobby Bus, its discoverer; Dan Britt, Kevin Cannon, Larry Nittler, Glenn MacPherson, Paul Steinhardt, Julie McGeoch and Malcolm McGeoch, Roger Fu, Alessondra Springmann, and Tim Elliott for meteorites and geology advice; and Cathy Plesko and Brent Barbee for planetary defense. I also thank John Brophy and Tom Prince for letting me participate in the Asteroid Retrieval Mission discussions; Nathan Strange, Damon Landau, and Marco Tantardini for celestial mechanics advice; Regan Dunn for information on dinosaurs;

Karen Daniels for granular materials advice; David Keith and Oliver Morton for an introduction to geoengineering; Daniel Faber, Chris Lewicki, Joel Sercel, Erika Ilves, Amara Graps, and Jim Keravala for insights into entrepreneurship; Joanne Gabrynowicz, Laura Montgomery, Rand Simberg, Bruce Mann, and Alissa Haddaji for information on the law of space. I received valuable help from Janne Robberstad, Kelly and Zach Wienersmith, Adam Dipert, and Tibor Balint regarding the arts in space; from Conevery Valencius, Kathryn Denning, and Michelle Hanlon on cultural issues; and from Sara Schechner on historical matters. In every case the individuals named supplied new understanding of the issues involved; any misunderstandings are mine alone.

There is a small cadre of space journalists whose articles are an invaluable "first draft of history" to whom I am truly indebted.

I thank Priyamvada Natarajan, for her support and for connecting me with Yale University Press; Joe Calamia, Jean Thomson Black, and Joyce Ippolito for their editorial assistance; and the Radcliffe Institute for Advanced Study for hosting our 2018 exploratory workshop on space resources.

I owe an unpayable debt to my engineering family from Birmingham, UK: my parents, Wilfred Hanson Elvis and Vera Helen Elvis, and my older brother Graham Elvis, who all assumed I would become an engineer too; finally, maybe, I have redeemed myself. And most of all, I thank my immediate family, Pepi Fabbiano and Camilla Elvis, for their knowledge of Latin, but mainly for their patience as I ranted about asteroid mining. Hopefully they will get some peace now.

A Note on the Text

Metric Measurement

Let's not propagate imperial or U.S. customary units into space! It would only make our lives harder. A simple mistake in imperial versus metric units led to the demise of a Mars probe, the *Mars Climate Orbiter*, in 1999. Only Myanmar, Liberia, and the United States do not use metric units.

It's not so hard. For kilometers think miles; for meters think yards; for metric tons think tons. In most cases it really won't matter, as the numbers involved with space tend to be much bigger than we are used to thinking about. (But if more precision is needed, then 1 kilometer is 0.6 miles; 1 meter is 3 feet, 3 inches; 1 metric ton is 1.1 U.S. tons.)

Special Terms

Several terms come up so often and are usually referred to by their acronyms or by their short-form names that it's helpful to list them for reference.

AU	astronomical unit, the mean distance of the Earth from the Sun
cis-lunar	within the Earth-Moon system
delta-v	change of velocity
ESA	European Space Agency
GEO	geostationary Earth orbit

ISS   International Space Station
JAXA   Japan Aerospace Exploration Agency
JPL   Jet Propulsion Laboratory
LEO   low Earth orbit
PPP   public-private partnership

Asteroids

Introduction: Why Boldly Go?

Why should we go to the asteroids? For that matter, why should we go to space at all? In the movies space is full of exotic planets and strange aliens. But the space we can actually get to, for all the beautiful pictures we see of Solar System worlds, seems lifeless, cold, and hostile. Is there room in the cosmos for us? Do humans have a future that spans many worlds? Or are we limited to Earth alone? That depends on how motivated we are. What would that motivation be?

In the science fiction TV and movie franchise *Star Trek* the crew of the starship *Enterprise* boldly go to explore strange new worlds where no one has gone before. Captain Kirk and his crew did it out of a sense of adventure. This very human urge to explore is often used as the main justification for going to space, and there is surely something to that idea—we do love the thought of setting out for the unknown. The problem is that decades of justifying space travel this way have not got us very far. In the course of 50 years only a few hundred people total have gone into orbit just above our atmosphere. Telescopes and robotic spacecraft have uncovered more and stranger new worlds, yet no people have set foot on these worlds. Why not?

Well, for one thing, space is expensive. And there's the rub. In the twenty-third-century setting of *Star Trek*, money is obsolete; anything you want is made in a device called a replicator. So the thrill of exploration is reason enough for the crew of the *Enterprise*. For us, though, choosing to put our tax dollars into pure exploration for its own sake is a real dilemma. There are so many other demands on those dollars. Is adventure travel for a few astronauts and armchair travel for the grounded public worth that price? In our early twenty-first-century world we have made the choice, and it was no. Boldly going into space just hasn't happened. Only a few people are now in space, and they are all on quite short visits.

Also, as the mantra goes, "Space is hard." Rocket science is unforgiving—a single small mistake can lead to total failure, a spectacular explosion. But other technologies were similarly precarious when they began. Sailing ships were often lost at sea, and steam engines in trains and ships frequently exploded in their early days. But that didn't stop us. Solving the technical problems is not the real barrier. "Space is hard" is true, yes, but it's also a way of giving up.

What we need is stronger motivation. If we are to see human exploration grow to match the scale of the Solar System, then we need a compelling reason to get off the comfortable couch that is the Earth.

Motives are vital; they are why we do anything. We are all called to action by a few powerful forces that get us moving. A trio of big motivators gets us to do a lot: *love, fear,* and *greed.*[1] These have, in turn, created immortal literature and art, raised great armies, and led us to the ends of the Earth in search of gold.

And these are the motives that will get us into space. Here's why:

- Our *love* of understanding our world—the need to know—draws us to scientific inquiry. The asteroids, as we will see, are deeply implicated in some really big questions.
- *Fear* of our own destruction, be it on a local scale or of the annihilation of the whole human species, tells us that we need to track down any killer asteroids out there that could hit our home planet.
- Our *greed* for the riches that are out in space, riches that could bring great benefits to the whole world, drives us to draw up a new map of the Solar System, a map on which many asteroids are marked with an X for "Here lies treasure."

The promise of all three motives holds our attention. They are the starting point of this book. That said, the great motivators don't always point in the same direction. Traders on the stock market, for example, are always balancing their greed for gain against their fear of loss. It is conflicts between our strong motivating forces that create great art. Literature is full of tales of love overcoming both fear and greed, but also of love failing in the face of those mighty forces. But I argue we can align these forces. I will show you that, when we think about going to the asteroids, love, fear, and greed all work together, pulling us out toward their realm.

As all the best whodunits say: motive is not enough; you also need means and opportunity.

The means involves bringing in skills spanning a wide range of talents: engineers to build spacecraft as well as a whole different set of engineers to build mining equipment; astronomers and geologists to prospect for the ore; businesspeople and economists to close the business case; lawyers to deal with the inevitable disputes, which may escalate to involve the need for policy experts, diplomats, and, though we hope not, the military.

Opportunity is here now thanks to the NewSpace movement, which is bringing a new commercial mindset to space. After 50 years and more of government-ordered, top-down planning, space exploration is now starting to operate like a business. NASA has encouraged this growth. As a result, space is wide open. Start-ups and legacy companies alike are pursuing a whole array of moneymaking ventures. Only a handful of successes are needed to create a new economy in space.

I believe that we will be making fortunes from space, and that fulfillment of our greed will also enable satisfaction of our love and fear. Greed is the motivation that was long ignored when it came to space. That has changed, as evidenced by those enthusiastic articles about the coming space trillionaires. Many of us "space cadets," the enthusiasts of space travel as a reward in itself, don't like to think in such mundane terms. We rather wish we were in the twenty-third century already. Nonetheless, greed is the one motive that will enable the others. Making a profit—preferably an obscene, gold rush–like profit—from space will produce a cascade of benefits for the science and security motives of going to space, the love and the fear. Once there are treasures to be had, space travel will pay for itself.

There are several ways besides asteroids that space could turn into a bonanza, but I'm putting my bet on the resources in the asteroids. Partly that's my bias because I'm an astronomer and astronomers can be really useful in getting that enterprise going. But it's also because the asteroids are the largest store of accessible materials in the Solar System and the one potential source of profit that gets us away from the smaller Earth-Moon system. They are the future.

Why is an astronomer beating the drum for asteroids and asteroid mining? After all, I've spent some happy decades using the best telescopes in the world to pursue my career in pure astrophysics research. I was trying, with some success, to understand the giant black holes at the centers of galaxies that light up so brightly when gas pours down toward them that we can see them all the way back to the earliest times of our universe. We call these things "quasars." (You'll find a few mentions of these exotica as I discuss the history of asteroids.) How did I get distracted by small lumps of rock in the foreground?

For me the starting point was wondering how future astronomers would be able to continue on the journey of exploration that I have been on for my whole career. This question, I realized, largely comes down to considering how NASA will be able to extend its long string of astoundingly successful astronomy missions, of which the telescopes *Hubble*, *Chandra*, and *Spitzer* have long been the flagships. Each one studies one band of wavelengths: optical, X-ray, or infrared. To study the giant black holes that shine brightly as the quasars I love, I rely on having a full complement of telescopes spanning all of these wavelength bands. That's because quasars don't respect our feeble technological limitations but instead spread their power democratically across the whole spectrum. "The stars are indifferent to astronomy," as the rock band Nada Surf says.[2] So I need telescopes that work in each wavelength if I'm going to understand them fully. And I'm not at all alone. Most astronomers now use multiple wavelengths to further their research, an approach that has been so productive that it has become almost a cliché to say that we live in a golden age of astronomy.

The problem is that these spacecraft-based observatories are getting much more expensive. So much so that the next generation of great telescopes will push the limits of what the U.S. Congress is willing to pay for. *Hubble*'s successor, the *James Webb Space Telescope*, costs some $9 billion. At that price we can afford only one. But we need a full set. Something must be done.

It's not just our exploration of the distant universe that's in trouble. We're not doing so well nearby either. The Solar System is big. In round numbers, a round-the-world voyage is 40,000 kilometers. Go 10 times that far and you could get to the Moon. But you'd have to go over 1,000 times as far to get to Mars, even when it's at its closest to Earth. Space is so big that we are far from exploring it at scale. Every 10 years NASA asks the National Academies to poll the planetary science community and come up with a report recommending the top priority destinations for the next decade. The latest result, the 2011 report *Vision and Voyages,* said that the top three priority large missions up to 2022 should be Mars, Europa, and Uranus.[3] Given the budget, all NASA can do is say, "Choose one." At that rate it will take a generation to do just these three. Other space agencies have similar problems. The European Space Agency (ESA) started to define its next goals in 2019. They are planning out to the year 2050, over 30 years away. Yet there are nearly 200 worlds for us to explore in our Solar System. I don't want to wait.

Arithmetic tells us that a great increase in funding and/or a huge lowering of costs is needed. The likelihood of the world's governmental space programs even doubling is small. The only alternative is to lower costs. Capitalism is a great tool for bringing down costs and increasing volume. If profits are made from space, those enterprises won't have to beg for budgets. True at-scale exploration could then succeed on the back of this hoped-for in-space economy.

It's not all unicorns and rainbows. I want human spaceflight, but I recognize that as an emotional yearning based on growing up with the Apollo program, *Star Trek,* and the movie *2001: A Space Odyssey.* But really we should get in touch with our inner Mr. Spock and be coldly analytical about space. As a scientist, I am trained to be skeptical. When boosters for space claim enormous potential profits, I instinctively put in the numbers, and they don't always add up. That hasn't made me popular with the would-be asteroid miners. In this book I take the same "enthusiastic but skeptical" approach. Throughout I try to separate the facts from the hype.

Why should we really be in space? What can we do there? How can the human enterprise in space grow into something important and good? I think the answer lies with the asteroids. They supply us with motive, means, and opportunity—and those motives are strong. They are love, fear, and greed.

PART

The Scene

1

Asteroids: A Primer

Asteroids are rocks in space. They can be smaller than footballs, or they can be mountains flying in space. A few are the size of small moons. Millions of them lie in orbits between those of Mars and Jupiter, a region called the Main Belt (figure 1).

A few asteroids get kicked out from the Main Belt to orbits that swing near the Earth. "Few" is astronomer-speak. We live with big numbers. There are actually about 20,000 of these "near-Earth asteroids" bigger than 100 meters diameter, about the size of a football stadium, and millions of smaller ones down to 20 meters across (mansion-sized). Fewer than 10 percent of these near-Earth asteroids have been found to date.

The asteroids in the Main Belt have been there, more or less, since the planets of the Solar System formed 4.5 billion years ago. Although the near-Earth asteroids formed at the same time, they stay in their near-Earth orbits only for a few million, maybe a few tens of millions, of years. To an astronomer, that's a brief encounter!

We see evidence of these passing asteroids with our naked eyes when we see shooting stars, which are officially called meteors. The dust from some asteroids (and from comets) enters the atmosphere and burns up as meteors. Larger pieces make it to the ground as meteorites. The world's largest collection of meteorites, over 20,000, is held at the Smithsonian's National Museum of Natural History. There are over twice as many spread over collections worldwide.

Clockwise from top left, Figure 1a, b, c, and d:

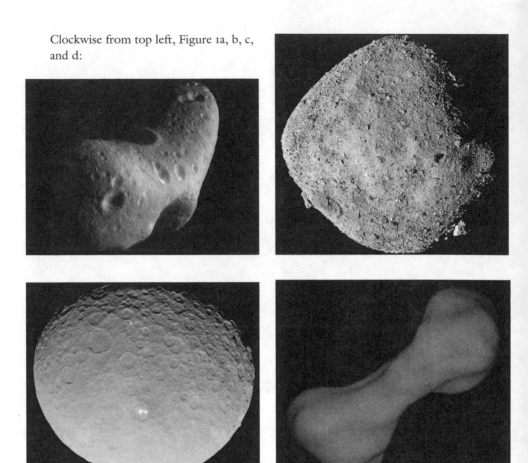

Figure 1: Asteroids come in a wide variety of shapes and sizes: (a) Eros was the first near-Earth asteroid discovered and is three times longer than it is wide (34 by 11.2 kilometers); (b) Bennu is the top-shaped asteroid, only 1/2 a kilometer across, visited by *OSIRIS-REx*; (c) dog-bone-shaped Kleopatra, imaged by radar, is 217 kilometers long; (d) spherical Ceres, visited by the NASA *Dawn* mission, is the largest asteroid at 939 kilometers in diameter.

Occasionally large pieces hit the Earth, such as the spectacular February 2013 breakup of a small asteroid over Chelyabinsk in Siberia. The Chelyabinsk event released as much energy as 33 Hiroshima atomic bombs. In the twentieth century there were several quite large impacts. The most spectacular was the 1908 Tunguska event that flattened hundreds of square miles of Siberian forest. These asteroids all had diameters of a few tens of meters, the size of a large house. This hazard makes near-Earth asteroids worrisome but also enticing, as they could be sites of possibly valuable resources.

Asteroids and meteorites come in a profusion of compositions. Francesca DeMeo, a planetary scientist at MIT, and her collaborators found 24 classification types for asteroids, and Tom Burbine, a meteoriticist (as scientists studying meteorites are called) at Mount Holyoke College, and his colleagues list 34 types of meteorites. Of these there are three main types: stony asteroids are mostly silicate rocks; carbonaceous asteroids are chockfull of tarlike carbon compounds; and metallic asteroids are nearly pure iron. The compositional profusion is due to their complex histories.

Why are these untidy asteroids out there? Asteroids are the "floor sweepings" left over after the main planets of the Solar System formed.

The eight major planets of the Solar System move around their orbits in grand isolation, not interfering with one another. This is not by chance. The planets formed from a giant disk of cold gas in which dust formed— the "pre-solar nebula." The *Hubble Space Telescope* has taken images of some of these "protoplanetary" dust disks around young stars in the constellation Orion, where they can be seen as shadows against the bright glow of the Orion Nebula (figure 2).

Disks are common in astronomy. Saturn's rings are a well-known example, but even giant black holes in the centers of distant galaxies have them. Gas clouds coming in on orbits from all directions would tend to collide and cancel out one another's motion, slowing down and falling inward. But gas clouds usually arrive around any big object with at least a slight preferential direction. When these gas clouds collide, they will still be moving, so they tend to merge. They form into a rotating disk.

The gas in this disk continues rotating unless some friction slows it down. Then it will be moving too slowly to resist gravity and will move inward. This slow inward motion is called accretion. Just what kind of friction in the gas removes angular momentum is an unsolved puzzle, though there are plenty of suggestions.

In the course of just a few million years—and again, that's a short time to an astronomer—the dust and gas of the disk clumped up and grew into

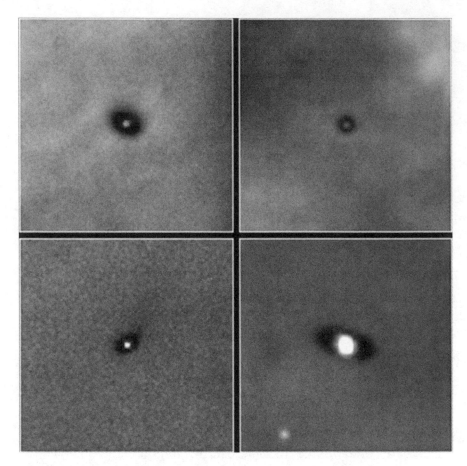

Figure 2: The dark oval shadows of the disk of planetary systems in formation are seen highlighted against the glow of a gaseous nebula in Orion in these *Hubble* images.

moon-sized bodies called planetesimals. We know how they get to be a meter across, but at that point they'll just break up again if they are held together only with gravity. Other, chemical forces must be keeping them bound together, but we are not sure just how.

The planetesimals were the progenitors of both the planets and the asteroids. They were quite small compared to planets, about 10 kilometers to 1,000 kilometers across, and possibly larger. At first there were many of these, strewn at random in a disk, all orbiting around the Sun. Naturally, they had collisions, and either they combined into larger and larger globes or they were split apart. Eventually, all these collisions led to a few large planets: each one cleaning out the debris from a ring around its orbit—either annexing the debris or shooting it out with a gravitational slingshot, depending on how the planetesimal encountered the debris. This had to be the outcome. If two planets had been still close enough to pull on each other strongly by gravity, they just would not have stayed around for 4.5 billion years, as the major planets have done. Instead one would likely have been thrown out of the Solar System. Probably several were. This weeding-out process is how the major planets got selected. That's why they are so far apart. As a result it takes a long time to travel between them.

But there is one place not so far from us where this planet-forming process stalled. Outside the orbit of Mars there is a region where Jupiter, the largest planet, stopped the planetesimals from merging. The gravity of Jupiter kept perturbing the orbits of these dust clumps, stopping them from combining into a planet. Instead they continued to bang together at high speeds. Most of them were broken up into smaller and smaller pieces. In doing so they formed the asteroids. That's why the Main Belt lies between Mars and Jupiter.

The history of the planetesimals explains the three main types of meteorite we find: the metallic, the stony, and the carbonaceous types. A large planetesimal naturally had three layers, rather like the Earth with its core, mantle, and crust.

Radioactive heating, coming mostly from the decay of an isotope of aluminum into one of magnesium, melted the large planetesimals. After another few million years, the planetesimal cooled enough that crystals began to form. The heavier pieces settled toward their cores. But not everything crystallized. Iron in particular was incompatible, meaning it didn't crystallize out of the molten material, the melt. The melt got very heavy with all that iron in it, and so it sank and formed a metallic core to the planetesimal. Something similar happens when seawater freezes. The salt-rich water is expelled from the ice and sinks. The Earth has an iron core for the same reason as the planetesimals. These cores are more accurately nickel-iron, as iron and nickel are the two most common of the heavy elements.

Outside the metallic core a stony layer forms, made up of all the silicate crystals ejected from the melt, and where heating has destroyed or driven off all complex molecules. Because the prime constituent of rocks is silicon, and silicon is 23 times as common as iron in the universe but only one-third as dense, the silicate rocks form a much thicker layer than the core.[1]

Smaller planetesimals generated too little heat to melt, and so their rock stayed pretty much unchanged from its pristine primordial state. There is only 30 percent more carbon in the universe than iron, so the amount of carbonaceous rock is not so large.

The newly solid and the smaller, never molten, planetesimals then got broken up in giant collisions. The metal cores, now solid spheres of nearly pure iron, were themselves shattered and released into space. The exposed fragments of the core became the metallic asteroids and are almost pure blends of nickel and iron. The silicate layer gave us the stony asteroids and the small planetesimals gave us the carbonaceous asteroids. Because of the many high-speed collisions they have undergone, many asteroids end up as loose agglomerations of rocks held together only by their own weak gravity. They are called "rubble piles" by astronomers.

The iron cores were only nearly pure. When iron becomes molten other elements can dissolve into it, just as some materials dissolve in liquid water. They are called the siderophile, or iron-loving elements. (*Sideros* means "iron" in ancient Greek.) Gold dissolves in liquid iron, so it is concentrated in iron cores, including that of the Earth, making it much rarer in the Earth's crust than it would have been had this settling process never happened. Other precious metals, including platinum, are scarce for the same reason. They too are siderophiles and sank to the core of the planetesimals and of the Earth alike. That's why many metallic asteroids contain higher concentrations of precious metals than are found in terrestrial mines.

Some stony and carbonaceous asteroids contain small, almost spherical lumps of different material, just a few thousandths of a millimeter across. These "chondrules" have different compositions from the surrounding matrix of rock in which they lie. So this rock could never have been molten, or they would have mixed together. Chondrules were the first small rocks to form in the Solar System. Chondrules can be fairly rich in metals or the opposite. As they were never too hot, carbonaceous asteroids contain high levels of easily destroyed molecules, including complex organic compounds and water. From studying meteorites we know of dozens of minerals that are found only in asteroids because of the exotic conditions in which they formed.

How did we discover asteroids? The dawn of the nineteenth century was the time of the French Revolution and the Napoleonic Wars. In England Jane Austen was writing her novels, and in Vienna Beethoven was composing his symphonies. Interesting times, indeed. And it was an exciting time in astronomy, too.

Ceres, the first asteroid to be found, and still the largest, was discovered on the very first night of the nineteenth century, 1 January 1801. The discovery was not accidental. It was spurred by the discovery only 20 years earlier of Uranus by a German-born musician from Hanover, William Herschel, from his backyard in Bath, England. Uranus had slotted nicely into the mysterious Titius-Bode law proposed in 1772. This law pointed out that the distances of the planets from the Sun fit a neat arithmetical sequence. Predicting the location of Uranus seemed to mean that the proposition was true. But, frustratingly, there was a missing planet in the Titius-Bode scheme between Mars and Jupiter. One of the first international big science consortiums was put together in 1800 to search for this missing planet. The project leader was Franz Xaver von Zach, a Hungarian-born astronomer working in Gotha (in what is now central Germany). His "Celestial Police" team divided up the sky into patches to be searched by 24 astronomers across Europe.

Searching for a planet is not as vague a goal as it seems. All the planets move along one great circle in the sky because they lie in the flat plane of the protoplanetary disk of gas that they formed from. We call this the ecliptic plane because eclipses of the Sun happen only when the Moon lies in that plane. (They don't happen every month because the Moon's orbit is slightly inclined to the ecliptic plane.) Each of the Celestial Policemen had to search only a thin strip of sky.

It happened, though, that the astronomer Giuseppe Piazzi was then working in Palermo, Sicily. Piazzi had founded the Palermo Observatory just a decade earlier. Sicily now had the most southerly observatory in Europe, and this happened to put Ceres in a more favorable position for being discovered. Piazzi was not part of the Celestial Police but, chancing upon a new moving star-like thing and realizing it didn't quite fit the description of a comet, he thought it might be a new planet. Von Zach must have been crestfallen to hear of Piazzi's discovery, but he and his team didn't miss out entirely. Piazzi's observations continued only for a couple of months before he fell ill. By the time his data was published, in September, Ceres had moved in the sky to where it was too close to the Sun, from our point of view, to be seen against the brightened sky. How could it be found again? The mathematical genius Carl Friedrich Gauss, then 24 years old, invented a new orbit determination method just to refind Ceres. His "patched conics" technique worked on very little data. It was von Zach himself who recovered Ceres at the end of the year, close to where Gauss had predicted it would be in the sky. One day later Heinrich Olbers confirmed it. Ceres fit the description of the missing planet according to the Titius-Bode law. Surviving a test like that is a major triumph for any theory. As a result, Ceres was considered a planet for 50 years.

Awkwardly for the Titius-Bode law, just a year after Ceres was discovered (satisfyingly in the "right" place), another asteroid, Pallas, was found there too. It was discovered by the same Heinrich Olbers, observing from his house in Bremen (now in Germany). Olbers is well known to astronomers for pointing out (more than 20 years after he discovered Pallas) how odd it is that the night sky is dark. Realizing that something so obvious is, in fact, strange is a really hard thing to do. If the universe were infinite, Olbers argued, then every direction would have a star somewhere along it and the sky would be bright as the Sun's disk. It was a strong argument. Others had pointed it out before, it turns out, but Olbers brought it up at the right time. Olbers' paradox stumped astronomers for over a century. Now we understand that the universe did have a beginning, and that light can reach us only from the distance it can travel in the time since the beginning. That leaves lots of directions empty of stars, and so dark.

Olbers's finding of a second planet in the slot where just one was expected was almost as disconcerting. A rush of discoveries of asteroids with similar orbits followed over the next few years, and so astronomers began to number them consecutively: Juno in 1804 and Vesta in 1807 (also found by Olbers). The tidiness of the Titius-Bode law was clearly not working. As more and more objects like Ceres were found, astronomers had increasing misgivings about calling them planets. What were these things?

Soon after Pallas was discovered, Herschel, now far more famous as an astronomer than as a musician, questioned whether such entities were planets or perhaps comets, but concluded they fit into neither category of known Solar System bodies. Unlike planets, they did not appear as a disk; unlike comets, they did not show large tails. He concluded that they were something new. So they needed a new name. Herschel declared, "From this, their asteroidal appearance, if I may use that expression, therefore, I shall take my name, and call them Asteroids."[2] He took asteroid from the ancient Greek word *asteroeidēs,* meaning "star-like." For several decades they were still usually called planets, but by 1851 there were 15 of these new celestial bodies known. Clearly this was not the simple and orderly Solar System everyone expected. Gradually the term *asteroid* took over, and Ceres was dropped from the list of planets.

Confusing naming conventions are another professional hazard of astronomy. We must give names to the things we discover before we know what they are, or we won't be able to talk about them. Herschel himself cautiously said of asteroids that he was "reserving for myself, however, the liberty of changing that name, if another, more expressive of their nature should occur."[3] In my original line of work there are, for example, easily 20 terms for the supermassive black holes at the centers of galaxies when they are seen, lit up by gas pouring down upon them.[4] Each name applies to a different form of their discovery, and I'm to blame for a couple. After 50 years of hard work we can now see that they are all essentially the same. Mostly we now call them active galactic nuclei or, more simply, quasars. Coincidentally, *quasar* is short for quasi-stellar, which means star-like, exactly as *asteroid* does. (In case you were wondering, *asterisk* means "little star.") The asteroids caused the same terminological dithering among astronomers a century and a half before quasars.

Piazzi had died in 1826, so he did not witness the demotion of his discovery. But really, finding a whole new type of object is a bigger achievement in science than finding another of the same old, same old, even if it's a planet. That's especially true when the discovery upsets our whole idea of how something is put together. Finding the asteroids did just that. The terms *small Solar System bodies* and *minor planets* came to be used interchangeably to refer to asteroids, comets, and other nonplanetary bodies, such as the Trojan asteroids found in special stable locations ahead and behind the major planets in their orbits, especially the massive Jupiter. Altogether, they show that the Solar System is much more complex than we had thought.

The changing classification of Ceres is very like Pluto being kicked out of the canonical list of Solar System planets. Pluto too was considered a planet for 50 years, but again, astronomers had increasing misgivings as more and more objects like it were found. Something had to change. It was a poor public relations job by the International Astronomical Union to unceremoniously eject Pluto from the list of planets. It should have said that Pluto was getting a promotion from "runt of the planets" to first of a new type of celestial body—the dwarf planet. In fact, when the International Astronomical Union voted to put Pluto in this new dwarf planet category in 2006, it also placed Ceres in there. So now Ceres is both an asteroid and a dwarf planet! No one ever said that astronomy is overly tidy.

How do we know there is a wide range of compositions for asteroids? They appear only as points of light moving in the sky. We can see what meteorites are in detail by analyzing them in the laboratory. That doesn't guarantee that asteroids are the same, does it? The answer is that we find out using spectroscopy, which is the tool that changed astronomy from a largely mathematical science of measuring positions on the sky into astrophysics, the study of how the universe formed and evolved.

Asking what the stars were made of, even though it's an obvious question, was left alone for centuries because no one had the slightest clue how to begin to answer it. It took advances in both optics and chemistry to provide the tools that were needed.

Isaac Newton split sunlight into a spectrum using glass prisms in the late seventeenth century. He showed, to everyone's great surprise, that white sunlight contained all the colors of the rainbow. It wasn't until over a century later, though, in 1802—the same year that Olbers discovered Pallas—that William Hyde Wollaston found there were dark bands in the spectrum of the Sun, and that they always came at the same locations. All the colors of the rainbow aren't, it turns out, all the colors that there are. Sunlight is missing a few. What were the dark bands due to? A big clue came just two years later. Thomas Young, an English polymath who also helped decipher the Rosetta Stone, showed that light is a wave, and that each color of light corresponds to a different wavelength. Blue light turns out to have a shorter wavelength than red light.

It was 10 years later, a year before Napoleon's final defeat, when the crucial breakthrough was made. In 1814, the Bavarian Joseph von Fraunhofer invented the spectroscope. Fraunhofer was a glassmaker who was constantly searching for new ways to make better glass for his company. So he was keen to measure the index of refraction—how much the glass bent light through an angle—of his experimental new glasses. Fraunhofer replaced Newton's prisms with his new invention of a diffraction grating, which is a sheet of glass with very fine parallel lines drawn on it. The lines are separated only by about the wavelength of visible light, and this closeness makes them spread out, or diffract, light that shines onto them. It spreads the light out far more than a prism can. Where before we could make out seven colors traditionally, and maybe a few more in reality, a diffraction grating let us see hundreds or more. The more detailed look at the colors enabled by diffraction gratings revealed that Wollaston's bands split up into sharp dark features at very specific wavelengths. No one at the time knew why these wavelengths were special. Fraunhofer didn't care. These sharp lines, whatever they were due to, were just what he needed to calibrate his gratings and prisms accurately, and that helped his optics business stay ahead of the competition. His diffraction gratings lie behind a vast number of modern astrophysics discoveries.

The nature of Fraunhofer's lines remained mysterious for another 45 years. Only then did Gustav Kirchhoff show that the Fraunhofer lines in the Sun were at exactly the same wavelengths as those produced by common elements heated in the laboratory. It turns out that when different elements are thrown into a flame, each one produces a quite distinct pattern of sharp lines in its spectra. These are so distinct that Kirchhoff's collaborator, Robert Bunsen, realized that there were two new elements—he called them cesium and rubidium—just from observing their spectral patterns. By matching the patterns for the elements to Fraunhofer's dark lines, Kirchhoff could see what the Sun was made of!

There's a possibly apocryphal tale about how he came to do this. One day Kirchhoff and Bunsen saw a large fire in the nearby city of Mannheim from their laboratory in Heidelberg. So they used their spectrograph to determine what was burning. Soon after the fire, so it is said, Kirchhoff and Bunsen were taking their exercise along the scenic Philosopher's Walk in Heidelberg talking about this feat.[5] Bunsen half joked, "Why should we not do the same with regard to the Sun?" Moment of silence . . . *Bingo!* They had a blinding glimpse of the obvious—if they could see what was burning in a fire 20 kilometers away, then there was no reason they could not ask what the Sun, and even other stars, are made of, though they be millions of kilometers away. Kirchhoff and Bunsen published their paper on the chemical composition of the Sun the very same year, in 1859.[6] Modern astrophysics was born.

As for elements, so for the combinations of elements we call minerals. When sunlight falls onto a rock, including the rocks that asteroids are made of, they reflect some of that light. How do we know which one we are looking at? Different rocks have different colors. That is, they reflect better at some wavelengths than others, depending on what they are made of. The three main types of asteroid have distinctive spectra. The metal ones reflect almost all colors of light the same, so their reflected spectrum is almost featureless. Normal silicate rocks don't reflect too well near the red end of the optical spectrum (at a wavelength of about 1 micron, one-thousandth of a millimeter). Water or, more accurately, water bound into rocks such as clays doesn't reflect so well in the middle of the optical spectrum (at about 0.7 microns wavelength). So if we can take a spectrum of an asteroid and compare it to the Sun's spectrum to see at what wavelengths the asteroid is reflecting most and least, we can tell what it is made of. This gives us only a rough idea of asteroids' composition, as solid minerals don't have the sharp lines in their spectra that gases do. Still, asteroid spectra are good enough to say which of the three main types they belong to, at least most of the time.

How much sunlight they reflect is called the albedo. (*Albedo* means "whiteness" in Latin.) The type of rock gives us some idea of the asteroid's albedo. The three main types of asteroid—stony, carbonaceous, and metallic—reflect very different amounts of sunlight. Metallic asteroids can be shiny, reflecting a quarter of the sunlight falling on them; carbonaceous asteroids instead reflect just a few percent, about the same as fresh asphalt. That's 10 times less than metallic ones. If we assumed a carbonaceous asteroid was metallic, we'd think it was several times smaller than it actually is.

We have another tool to use, thanks to another discovery by William Herschel made just a couple of years before he coined the term *asteroids*. This method can give us a much better measurement of how big an asteroid is. Herschel wondered if there might be light beyond the colors we could see. He knew that a thermometer in sunlight registers a higher temperature than in the shade. So he put a thermometer beyond the red end of the spectrum of the Sun made by a prism. There was nothing he could see there—yet, despite this, the temperature of the thermometer went up. There had to be some invisible radiation beyond the color red that was warming the thermometer. Hence it was called infrared—beyond the red. Probably for this reason, infrared radiation was misleadingly called heat rays for a while, notably in *Superman* comics. Similarly, heat lamps are bright infrared emitters. Infrared radiation is important in studying asteroids because it gives us a new way to measure how large they are. At wavelengths of around 10 microns, 10 times longer than any light we can see with our own eyes, we reach the brightest part of the spectrum of a near-Earth asteroid. If we could get an infrared spectrum of every asteroid to check on that emission, we'd know just how big the asteroid is.

That's because of a useful idea in physics called a black body. This term, invented by Kirchhoff the very next year after discovering how to learn the composition of the stars, refers to an ideal object that reflects no light at all; it is perfectly black, with zero albedo.[7] No real material is that non-reflective. But it turns out to be an incredibly useful tool in physics and astronomy. That's because, although a black body does not reflect light, it does radiate light. Everything radiates, including you, and the hotter something is the more it radiates. We are familiar with this from seeing iron glow red-hot when it is heated enough. The spectrum of this radiation has a special shape that we call a black body spectrum. Hotter objects emit shorter wavelength radiation in a very predictable way. So the blue of a gas flame is hotter than the glowing red iron it is heating.

The most perfect naturally occurring black body known is the cosmic microwave background. This background is the remnant glow from just after the Big Bang. Princeton physicist Bob Dicke was looking for it in 1965 with a specially built radio telescope. But he was scooped by Arno Penzias and Bob Wilson, who were at the nearby Bell Labs in Holmdel, New Jersey. Penzias and Wilson got the Nobel Prize for that discovery. But they measured only one point, so the talk announcing that the spectrum of this cosmic background had *exactly* the shape of a black body got a (very rare) standing ovation for John Mather of NASA at the American Astronomical Society in 1992. (Unfortunately I had to be in a parallel session in a small room next door; we just heard all the cheers through the thin partition.) That measurement earned him a Nobel Prize. That black body had a temperature not quite 3 degrees above absolute zero.

It turns out that we can use black body radiation to measure the size of asteroids. Most near-Earth asteroids, being about the same distance from the Sun as the Earth (1 astronomical unit), receive about the same amount of radiation as the Earth on each patch of their surface, so they warm up to about the same temperature. (The Earth's atmosphere complicates things so they are not identical in temperature.) They then reradiate their absorbed solar energy as a black body. On the absolute Kelvin temperature scale, this is about 300 degrees. (The Kelvin scale is the same as Celsius, except that it begins at absolute zero rather than the freezing point of water. Zero degrees Kelvin turns out to be minus 273 degrees Celsius. The Kelvin scale is universally used in astronomy, although Celsius is used by meteoriticists.) On the Kelvin scale calculations become easier. The formula for the wavelength where the black body radiation peaks is simple on the Kelvin scale, as found by Wilhelm Wien in 1893: a black body at half the temperature is brightest at double the wavelength. The surface temperature of the Sun is about 6,000 Kelvin. (6,000 Kelvin is way hotter than your oven, which can get to just about 530 Kelvin, but it's not especially hot for astrophysics.) A 6,000 Kelvin black body peaks at about 0.6 microns, in the middle of the optical band. It's no coincidence that our eyes are sensitive to the brightest light available from the Sun, though if our atmosphere were not transparent there, we would presumably use whatever wavelengths did get through.

Professor Wien's law then tells us that a black body 10 times cooler than the Sun peaks at a wavelength 10 times longer, at 6 microns, and a 20 times cooler black body of 300 Kelvin, like a near-Earth asteroid, will peak at 12 microns. These wavelengths are often called the thermal infrared. (The name arises because bodies at common Earthly temperatures, about 300 Kelvin, emit the peak of their radiation there.) This is a temperature far cooler than most stars, galaxies, or other astronomical bodies, like the Sun. So in pictures of the sky taken in this band near-Earth asteroids stand out clearly.

Given a temperature, how bright an object appears to be in its black body emission depends only on the area we see, and how far away it is. Since the orbit of the asteroid tells us its distance from us, once we have the temperature and brightness, we can work out its size. Even better, comparing that size with how bright the same asteroid appears in visible light tells us how good a mirror the asteroid is. That is: what fraction of the sunlight hitting it is reflected? This is a more formal way of describing the albedo of the asteroid. The full calculation is a bit more complicated due to the "thermal inertia" of the asteroid surface. Thermal inertia just describes how long the asteroid surface continues to radiate away its heat even when it rotates out of direct sunlight. In just the same way, a sidewalk has thermal inertia because it is still hot on a summer night, even though there is no more sunlight heating it up.

It's a pity, but Main Belt asteroids are hard to get to. It takes a lot of energy and time to reach them. With our current rockets, round-trip journey times from Earth take a decade or more. NASA's *Dawn* mission took nearly four years to reach the large asteroid Vesta. Hard-to-reach places really put a damper on our motivation for going there, whether it is love, fear, or greed.

Luckily Jupiter has given us the near-Earth asteroids. If an asteroid is unlucky enough to stray to the wrong place, Jupiter's gravity can move it out of the Main Belt, often toward the inner Solar System. That's great for us, because these orbits make them easier to reach with our rockets. Once an asteroid is a near-Earth asteroid it stays on that handy orbit for a few million years, but then is likely to be scattered off again. Its future is either ice or fire, as it is either flung off into the cold of the outer Solar System or dives in toward the heat of the Sun.

How Jupiter does this is a bit subtle. It's a result of the Yarkovsky effect. Ivan Yarkovsky was a Polish-Russian civil engineer who also published scientific papers. Around 1900, when he was in his late fifties, and only two years before he died, he self-published a pamphlet about how a spinning asteroid would behave. His work was forgotten for decades, partly because he was trying to defend the ether theory that was being put to rest by Einstein and others around the same time. But he was right about his effect, and his long-lost pamphlet was recalled in 1951 by Estonian astronomer

Ernst Opik and developed into a useful theory. Opik had read the pamphlet in his native Estonia in 1909, long before he fled the advancing Red Army in 1944. Opik published his study in 1951, and Viktor Radzievskii published the idea independently in the Soviet Union a year later.[8] (A careful hunt by George Beekman located Yarkovsky's original pamphlet in 2003, in a library in Moscow.)

Yarkovsky realized that an asteroid is heated on its noonday side by the Sun but takes some time to cool off due to its thermal inertia, and in this time it rotates. So the heat the asteroid emits goes off in a different direction from where it was absorbed. This produces a tiny force that pushes the asteroid either a little faster in its orbit or a little slower, depending on which way it is rotating. This is just like firing a small rocket. A retro thrust lowers the orbit, and a positive thrust raises the orbit. The force is so small that it takes millions of years to make a major difference to the orbit. But the Solar System has plenty of time.

Gradually the Yarkovsky effect makes some Main Belt asteroids' orbits drift into resonances with Jupiter. That means that an asteroid's orbit is timed just right so that Jupiter always has its largest pull on it at the same place in its orbit. Resonances are familiar things with an unfamiliar name. When you push your child on a swing you time the push to happen at just the right moment so the swing goes higher. In the cases of asteroids, if their orbit is timed right they get an extra push and their orbit moves out more and more until the asteroid is moving on a highly elliptical orbit. Some of these orbits eventually reach far in toward the Sun and get close to, or even cross, the Earth's orbit.

A near-Earth asteroid is any asteroid that strays within Mars's orbit at least some of the time. That means that its closest approach to the Sun (or perihelion, called q) is less than the radius of Mars's orbit, which is 1.38 astronomical units (or AU—the average distance between the Earth and the Sun) at its smallest value, its perihelion. Mars has a more elliptical orbit than the Earth and its furthest point from the Sun, its aphelion, is 1.66 AU.

The scattering that near-Earth asteroids undergo is quite a random event and can throw them far out of the plane in which the major planets lie. (This is the ecliptic plane that Piazzi and the Celestial Police searched.) As a result their orbits are often highly tilted compared to the Earth's. That makes it energetically more expensive to get to them. Even a modestly inclined orbit at 20 degrees can take more than 10 times the energy to reach.[9] For now, at least, we are limited to visiting just the ones whose orbits are only slightly inclined to the Earth's orbit.

Over 20,000 near-Earth asteroids have been identified and cataloged, primarily from ground-based telescopes surveying in the optical band. The largest is Ganymed, which is about 50 kilometers across. (Ganymed should not be confused with Ganymede, the largest moon of Jupiter; but it is.) Most are much smaller than Ganymed, down to the size of a house, or even less. That's at least partly because the Yarkovsky effect is larger on smaller bodies. Twenty thousand asteroids sounds like a lot, but our surveys are incomplete, woefully so for smaller ones. But our catalog is growing pretty fast. About 2,000 near-Earth asteroids are now being discovered every year. New discoveries are trending toward finding smaller and smaller near-Earth asteroids as the searches become more sensitive, and also because we have found most of the larger ones.

Large numbers of really small asteroids hit the Earth harmlessly every year, and fragments of them reach the ground intact. The small asteroids (smaller than a meter across diameter) are called meteoroids. As they go through the atmosphere they make brighter tracks than meteors. These are called fireballs or, in extreme cases, bolides. The fragments that reach the ground are called meteorites. These meteorites are valuable messengers from the early times of the Solar System.

Until recently I had never studied any meteorites up close. Frankly, they didn't seem that interesting. Then I got the chance to participate in an introductory meteoritics class given by Professor Roger Fu in Harvard's Earth and planetary sciences department. He passed around a dozen meteorites for us to classify. A bit of guidance from an expert about what you are looking at helps a lot, focusing you on the things that matter. The first thing Roger pointed out to us is that we were holding in our hands rocks that are billions of years older than any rocks on Earth. Most rocks on Earth are barely a tenth that old. Meteorites are as old as the Solar System. Touching them is pretty awesome. These rocks began to seem a little strange and exotic after all.

Then we looked at the meteorites and gauged their density simply by hefting them in our hands; practiced geologists can get pretty good at this.

After we'd written down our notes on each one, Roger had us look for trends—what property went with what? The metallic ones were the densest; they are nearly pure iron, after all. The ones with small round lumps, called chondrules, were next, while the stony ones without chondrules were the lightest. "Why is that so?" Roger asked us. Embarrassingly, none of us got the answer. Yet the answer is really quite simple. The ones with chondrules have never melted, so they still have the solar composition of elements—minus the light elements hydrogen and helium—that is universal in astronomy, including iron. This mix of elements is set by the nuclear physics happening in stars, so it never changes. But when heating melts the rocks of a planetesimal, including the chondrules, then the elements can move and separate out. The heavy iron separates out, leaving the remaining stony part lighter than it was originally.

The earliest iron worked in both Egypt and in pre–European contact Greenland came from meteorites. Why Greenland? Because a dark iron meteorite stands out clearly against the white of the snowpack, and Greenlanders really had no other choices for metals. The iron used in Greenland seems to have come from the Cape York meteorite.

Why would the Egyptians use meteorite iron? For one thing, dark meteorites are also easy to find in the desert. But another good reason is because an iron meteorite is already virtually pure iron. So getting started is easy. There's no need for high-temperature smelting to separate out impurities. Also, meteoric iron is unusually rich in nickel. Nickel-iron is a form of steel. So meteoritic iron is actually better than iron dug from the Earth; it resists rusting. Did the Egyptians know that iron they used had fallen from the sky? Maybe not at first, but at some point, they did. An Egyptian hieroglyph, *bia-n-pet*, literally meaning "iron from the sky," appeared around 1295 BCE.[10] Perhaps in that year there was a spectacular fall of meteorites, or maybe a large impact that made it clear where the iron came from.

Modern Europeans and Americans had a hard time believing that rocks could fall from the sky. It was a reasonable skepticism. It seemed much more likely that meteorites came from volcanoes than out of thin air. Earlier well-witnessed reports, such as the 127-kilogram rock that landed near Ensisheim, Alsace, in 1492, may not have been widely known—or believed.[11] But in that same rush of discoveries around 1800 that brought the discovery of asteroids, the case for stones falling from the sky became hard to deny.

In 1793 a lawyer, Ernst Chladni, had a conversation with Georg Lichtenberg, a physics professor who had witnessed a fireball and wondered if it had been some cosmic body entering the atmosphere. Chladni decided to find out. He spent three weeks compiling all the accounts of fireballs in the sky and rocks falling to Earth that he could find in the university library in Wittenberg (now in Germany). He applied his lawyer's training to evaluate these eyewitness accounts. He found two dozen, and they were all highly consistent, though they were widely separated in location and time. He then used the accounts to estimate how fast the fireballs were moving and came up with a speed much higher than could arise on Earth. He published his results the next year. This was all too novel, though, and too much based on indirect evidence, for most scientists at the time.

As luck would have it, the very next year a large meteorite fell near a cottage in Yorkshire, England. Two chemists, one British and one French, analyzed its composition and found a large amount of nickel in the iron of the meteorite. Such an alloy had never been found on Earth, so they suggested it was of extraterrestrial origin. They also discovered chondrules, something never present in terrestrial rocks. They published their results rather late, in 1802, just a year after Ceres was discovered. The clinching evidence came just a year after that. A bright fireball was followed by thousands of rocks falling near L'Aigle, Normandy, in 1803. Jean-Baptiste Biot, a well-regarded French scientist, analyzed the distribution of meteorites on the ground and found that they lined up with the path of the fireball. He concluded these meteorites were indeed coming from space. His work convinced scientists throughout Europe.

There's a famous (if dubious) report that Thomas Jefferson, when president of the United States, was not so convinced. This despite careful investigations by two Yale professors, Benjamin Silliman and James Kingsley, of a meteorite that fell in Weston, Connecticut, in 1807. Jefferson is supposed to have said, "Gentlemen, I would rather believe that two Yankee professors would lie than believe that stones fall from heaven." Cathryn Prince, a historian who has investigated this claim carefully, suggests that the "quote" is not real, but that it may have started with a speech by

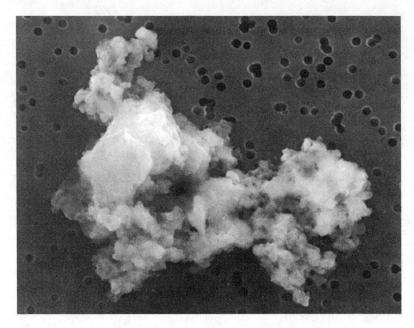

Figure 3: A tiny interplanetary dust grain, about 10 microns across, 1/5 the diameter of a human hair. The grain was collected in the stratosphere by a U2 aircraft.

Silliman's son.[12] She documents an antipathy between the Yankee Silliman and the southerner Jefferson that may have led to a wish to put down Jefferson's otherwise strong scientific reputation.

Recognizing that meteorites came from space gave early nineteenth-century scientists a way to study what asteroids are made of, just as spectroscopy gave them ways to understand what stars are made of 50 years later.

Meteorites can be found anywhere, and by now there are over 45,000 meteorite samples in collections. Antarctica is one of the best places to find meteorites. Just as in Greenland, the snowpack and absence of terrestrial rocks near the surface make meteorites easy to spot. The more than 20,000 meteorites collected by the U.S. Antarctic Search for Meteorites program have all been classified by the Smithsonian's National Museum of Natural History in Washington, DC, where they are stored. The Japanese pioneered Antarctic meteorite hunting, and a similar number are in the National Institute of Polar Research in Tokyo. Other large collections are to be

found in the U.K. Natural History Museum, Canada's National Meteorite Collection in Ottawa, and NASA's Johnson Space Center in Houston.
Many universities have large collections too, and several super-keen individuals have large collections, such as Naveen Jain, a billionaire entrepreneur, who is backing Moon Express, a space start-up. Jain has 500
specimens, all of them "falls," meaning that they were collected soon after
they fell to Earth and so are in almost pristine condition.

There are also vast numbers of tiny particles, just a few microns or less
in size, too small to be described as meteorites. These are interplanetary
dust particles (figure 3). Many of them come from comets. Both in these
dust particles and in meteorites we find pre-solar grains. They are rare and
make up no more than 1 part per 10,000 of meteorites, and up to about 1
percent of the interplanetary dust particles. As their name suggests, pre-
solar grains are older than the Solar System. Some were formed in supernovae, the explosions of massive stars, others in the gentler winds blowing
off the surface of old red giant stars. Larry Nittler works at the Carnegie
Institution for Science, where he specializes in what a detailed analysis of
meteorites tells us about the formation of the Solar System. Larry can tell
you how he picks out these tiny grains of dust from meteorites. These
pre-solar grains have a tortured history. As he and his collaborator Nicholas Dauphas state, these pre-solar grains "formed in stellar outflows more
than 4.6 [billion years] ago, became part of the Sun's parent molecular
cloud, survived formation of the Solar System, and became trapped in
asteroids and comets, samples of which now intersect Earth as meteorites."[13]
It's remarkable that any survived for us to find. They link us to the cosmos.

PART

Motive

2

Love

In English *love* is a very broad term, covering romantic love, love for your kids, your parents (if your parents are lucky), for football teams, and much, much more. The love I'm talking about is the love of knowledge, the love of finding things out. We are all curious. Scientists are lucky—we get paid to be curious.

How basic a motive is love of knowledge, though? Knowing more about your surroundings than your possible predators, including human ones, seems like a pretty basic motivation for staying alive. "Exploration is an important survival strategy in evolution," says R. B. Setlow, a senior biophysicist emeritus at Brookhaven National Laboratory.[1] It is often said that humans have an innate love of exploration. A stirring statement of this innate love was made by Stewart Weaver, a mountain climber and professor of history at the University of Rochester: "For all the different forms it takes in different historical periods, for all the worthy and unworthy motives that lie behind it, exploration—travel for the sake of discovery and adventure . . . seems a human compulsion, a human obsession even (as the paleontologist Maeve Leakey says); it is a defining element of a distinctly human identity, and it will never rest at any frontier, whether terrestrial or extra-terrestrial."[2]

Polynesian expansion eastward across the Pacific is perhaps the purest example of exploration. Ben Finney, a super-cool professor who worked at the University of Hawaii—he studied surfing for his master's thesis— was a major mover behind the rebirth of Polynesian sailing techniques in

Hawaii. For about 300 years the Polynesians expanded across the Pacific far faster than population pressure could account for. A tradition of primogeniture may have provided a push for younger sons to set out voyaging, but presumably they just wanted to see what was over the horizon. Finney compared their voyages to space travel: they sailed upwind, just as we have to push against gravity; and they had to take all their supplies with them since they really were going to desert islands, where no one had gone before, just as we do in space.[3] In one way we have it easier: we know where we are going.

Is the drive to explore enough to take us to the asteroids? As I've described them, they are just rocks in space. Geologists love rocks. But why should the rest of us care? Why should asteroids inspire our love of exploration? The answer is that asteroids are much more than "mere" rocks. They are implicated in some Big Questions. Some of the biggest there are, in fact. Where do we come from? What are we? Where are we going? These are questions we all ponder at some point.

The asteroids mainly help us with that first question, our origins. They are helping us answer a series of questions that lead us, one by one, to our civilization. How did the Solar System and our home world form? Where did the oceans come from? How did life start? Why are there rich mines? Mines matter because without easily mineable iron, our technology would likely have stalled out at a much simpler level. That makes this last one an important question. "Why are there mines?" is another of those questions like "Why is the night sky dark?" that, until you think about it, don't seem like puzzles at all. Let's look at how asteroids illuminate all of these questions.

Origin of the Solar System and the Earth

Most asteroid scientists are driven to understand how our Solar System came to have its present seemingly simple form. That's a pretty grand goal.

Our Solar System has a clear division between the inner planets and the outer ones. The four small rocky "terrestrial" planets lie in the inner Solar System (Mercury, Venus, Earth, and Mars), within 2 AU (the Earth being at 1 AU). In the outer Solar System (beyond 4 AU) come very different planets: the two gas giants, Jupiter and Saturn, and then the two, much smaller ice giants (Uranus and Neptune). For several decades there was a (relatively) simple textbook theory of how this simple two-zone ordering of the planets came about.[4] The idea is simple: find the snow line.

The disk of the pre-solar nebula of dust and gas was hotter toward the center. Materials that vaporize at low temperatures, especially organic molecules rich in carbon and hydrogen, are destroyed at low temperatures. They are called volatiles and are only found quite far out. Materials like carborundum that vaporize at high temperatures are instead dubbed refractory materials. Only these tough refractory materials could form out of the hot gas of the proto-planetary nebula close in toward the Sun. These materials formed the terrestrial planets. Moving outward, the temperature of the gas making up the nebula was cooler, crossing zero degrees Celsius at a particular distance out. The distance from the Sun where this happens is called the snow line (also called the ice line or the frost line). At the time the planets formed, just a few million years after the Sun began to shine, the line lay at about 4 AU. That is just outside the asteroid Main Belt and inside the orbit of Jupiter. As outside of the snow line the pre-solar nebula had a lot of hydrogen, carbon, and oxygen, those planets could scoop up large amounts and grow into gas giants. The pre-solar nebula became more tenuous at larger distances. Jupiter lay just outside the snow line, so it was the largest, and the other planets were smaller because there was less raw material from which they could grow.

All is explained. Beautiful. Neat. But, we now know, woefully incomplete.

There has been a huge revolution in planetary science in the past 20 years. This field used to suffer from having only one example, our Solar System. (Cosmology has the same problem, because we can observe only one universe.) Everyone was trying to predict how planets formed, and everyone assumed that our Solar System was normal. This is a good "Copernican" approach, which says we are nothing special in the universe, and it produced a good model that worked pretty well. But how could we know if it was correct unless we had another example to test it against? Maybe there was some other way our arrangement of planets could have happened that we hadn't thought of? Indeed there was.

Searching for planets outside our Solar System was considered a joke when I started in astronomy in the 1970s. First off, it was self-evidently impossible to find them, far outpacing the technology we had. Second, aren't planets a bit parochial? Astronomers were off discovering extraordinary new things far beyond our imaginings: pulsars, quasars, dark matter, gamma-ray bursts, and more. How could gas giants and small rocks compete? Luckily, we were all wrong.

Despite the prevailing skepticism, some astronomers were able to discover planets outside of our Solar System: exoplanets, as they soon came to be called. This was possible because of the sociology of astronomy. Astronomy is lucky to be a many-stranded field of research. There are dozens of large telescopes in the world, and they are run mostly independently of one another. As a result, though there are fashions, no one group or approach dominates. The loosely linked nature of astronomy has long insulated it from the groupthink that tends to spread when a field is driven to unite around a single project. Usually this uniformity of opinion comes about because the next step becomes too expensive to have many experiments running in parallel.

The first really convincing planet circling a normal star other than the Sun was 51 Pegasi b. Its discovery was announced by Michel Mayor and Didier Queloz, both of the University of Geneva, in 1995. Their announcement caused shock waves among astronomers. Any exoplanet would have been a major discovery. But this one was a huge surprise. It was a large planet, which had to be a Jupiter-like gas giant; yet it orbited its star, 51 Pegasi, in just over four days, putting it 20 times closer to 51 Pegasi than we are to the Sun! Clearly our Solar System was not the only kind.

Now we know there are many more of these "hot Jupiters." They could not have formed where we found them, as they are deep inside the snow line, where it is far too hot for there to have been enough hydrogen from which to build them. Yet there they are. Hot Jupiters really are "strange new worlds," as they say on *Star Trek*. Mayor and Queloz were awarded the 2019 Nobel Prize in physics for their discovery.

The discovery of hot Jupiters changed everything. Even before the second hot Jupiter was found the theorists were abuzz. The answer is that gas giant planets do not have to stay where they formed; they can "migrate." A Jupiter-sized planet migrating so close to its star will cause havoc to any smaller planets or asteroids whose orbits it crosses. They are likely to get thrown out of that solar system, or they may even collide with the migrating Jupiter. So hot Jupiter systems are not likely places to find other Earths.

Migrating planets sound crazy, though. What would make a planet migrate? Early on, while there is still a disk of dusty gas around the star, the youthful Jupiters moving within would feel the "wind in their face" from that gas and get slowed down, dropping into orbits nearer the star. Luckily, nothing that extreme happened in our Solar System, as it would have destroyed the Earth (and Mars and Venus too) on its way inward. It may have been a close shave, though.

The idea of migrating planets led to a fresh look at our own Solar System. There's a promising, if still disputed, model of how Jupiter and Saturn both were migrating inward but, just as Jupiter got to Mars's orbit, they stopped, and eventually turned, moving outward again. This "Grand Tack" model says that they stopped because they cleared out most of the dusty gas in the disk that the planets were forming from, so they were no longer being slowed down by it. In the process they scattered vast numbers of planetes-imals into the outer Solar System and took away most of the disk material from which Mars was trying to form. Mars was left with only 10 percent of the mass it would otherwise have had. Had the giant planets continued their inward migration, Earth would have suffered the same fate.

Soon after this, just a few million years later, all of the gas in the disk had condensed into planets, moons, planetesimals, and dust. But that wasn't the end of the dynamic history of the Solar System.

The next stage is called the Nice model, after the city on the French Riviera where it was developed. The Nice model makes use of the small bodies of the Solar System—many of them forerunners of the asteroids—to move the giant planets. It was a big event when this model was published in 2005. It came out in three simultaneous papers in the prestigious jour-nal *Nature*.[5] Getting one paper published in *Nature* is considered a big success for any scientist. Getting three published *simultaneously* was pretty much unprecedented.

The four authors of the Nice model suggest that all those planetesimals formed in the dusty gas disk eventually managed to move both Jupiter and Saturn inward. Each individual planetesimal has only a tiny effect; they are only a millionth or less the mass of Jupiter, after all. But over a few hundred million years and many planetesimals the effect builds up. If most of the planetesimals get thrown inward, the planet moves outward, and vice versa. This process is reminiscent of the supposed Chinese proverb "A man who moves a mountain begins by carrying away small stones."[6]

Everything changes very gradually . . . until suddenly (in an astronomical sense) Saturn reaches a point where it circles the Sun once for every time Jupiter goes around twice. That makes the tugs that Jupiter exerts on Saturn add up on each successive orbit. The two planets have entered a resonance, the same process that creates near-Earth asteroids. The effect is to make Saturn's orbit more and more elongated. Saturn then crosses the orbits of many, many more planetesimals and flings any it encounters into wildly different orbits.

Big consequences come from this. Our neighborhood gets bombarded with many-kilometer-diameter rocks—planetesimals hurled by Saturn. That's why the face of the Moon is pockmarked with craters. This is just what the Nice model authors were after, a way to explain the "Late Heavy Bombardment" of asteroids into the inner Solar System, where they hit the Moon and the Earth. We'll hear more about how this bombardment may have changed the Earth in the next couple of sections.

Like any good scientific theory, though, the Nice model gives out much more than you put into it. It explains not just the Late Heavy Bombardment but lots of other odd features of the Solar System. Why does Jupiter have families of asteroids preceding and following in its orbit, the Trojan asteroids? Why are there so many moons in odd orbits, "irregular moons," around the outer planets? Where did the distant Kuiper Belt, with its dwarf planets like Pluto, come from? And, even more distant, why is there a vast cloud of comets, the Oort cloud, around our Solar System? The Nice model, refined over the years by its original discoverers and lots of other scientists, can explain all these diverse phenomena. Basically, all those scattered planetesimals had to go somewhere, and each of these space oddities is due to some of them going to each of these locales. It is not a universally accepted model, but careful, detailed calculations do seem to explain these features in decent quantitative detail.

Much of the stirring up of the Solar System by planetary migration leaves marks on the asteroids. The types of asteroid can reveal the history of their movements around the Solar System. Just how many there are of each kind, in what orbit, and the details of their geology—all these variables will pin down the chronology of our violent early history. They may support the Grand Tack and Nice models, change them fundamentally, or even show that they are just plain wrong. That's how science moves ahead: one idea implies a bunch of consequences, and that gives other scientists the chance to go out and see if the consequences can actually be found. Asteroid scientists are looking forward to doing just that by sending robotic spacecraft to several asteroids. A couple of such missions are under way, and more are planned. It will be an exciting period in Solar System exploration.

For many asteroid scientists this is enough. Understanding our Solar System's dramatic history is no small thing. But we have three more Big Questions, all about the Earth, that asteroids help us answer.

Origin of the Oceans

Why does the Earth have oceans? Our closest neighbor planets, Mars and Venus, have none. Yet oceans cover two-thirds of the Earth's surface, making Earth utterly unlike its dry siblings in the Solar System.

It's not that water is scarce in the universe. In fact, it may surprise you that water molecules are abundant in space.[7] The dark clouds in our Galaxy that are the nurseries of young stars contain enormous amounts of water. That is because the constituents of water, oxygen and hydrogen, are two of the most abundant elements in the universe. Hydrogen was formed soon after the Big Bang, when the universe became cool enough for electrons and protons to stay bound together by their electric charge. A hydrogen atom is the simplest atom: just one proton surrounded by just one electron. Oxygen was formed by nuclear reactions in the cores of massive stars and released into interstellar space when those massive stars exploded as supernovae.

So the problem isn't there being enough water to fill the oceans but keeping all that water inside the ice line and explaining how it could survive when the young Earth was molten. How could water survive that heat? The generally accepted answer is that it didn't. Earth was born bone dry.

One place in the Solar System where there was still a lot of water after the Earth cooled is out in the asteroids and comets. We just need to find out how to deliver oceanfuls of water from there to here. Back in the late 1600s Isaac Newton suggested that comets' tails replenished Earth's water, as documented by historian of science Sara Schechner.[8] His particular theory doesn't hold up. Comet tails would not provide enough water to fill the oceans. His basic insight, though, turns out to be correct. The early Solar System contained so much water in asteroids and comets that we now think they did fill the oceans of Earth. Once again the Late Heavy Bombardment that the Nice model comes into play. That rain of rocks onto the Earth and Moon would have brought along a lot of water with it.

How much is a lot when it comes to the oceans? The U.S. National Oceanic and Atmospheric Administration (NOAA) estimates that there are 1.3 billion cubic kilometers of water in the oceans.[9] That's 1.3 billion billion tons. That does indeed sound like a lot. How many planetesimals would it take to bring us that much water? We can use Ceres, the largest asteroid, and the best approximation to a planetesimal that we still have nearby, to make an estimate. Recent measurements by the NASA *Dawn* spacecraft indicate that as much as 30 percent of Ceres by mass may be water.[10] That comes to 1/4 billion billion tons. So, if none were spilled out into space in the collision, we'd need only about five Ceres-like asteroids to fill the oceans. That's a small enough number to show that the idea isn't crazy. Some experts think it is even likely.[11]

But the Late Heavy Bombardment would have pummeled Mars, Venus, and our Moon too. Where is their water? Lost to space. The Moon and Mars have too little atmosphere to prevent water on the ground there from evaporating. Venus instead has a thick atmosphere, but is so hot that water on the ground would boil away.

Seen from deep space our planet is a pale blue dot.[12] Why is the Earth blue? Because of our blue oceans. And the oceans are only blue because they reflect the color of the sky, the result of sunlight scattering in our atmosphere. On an overcast day, the ocean is gray. But only the parts of the oceans under a clear sky can be seen from space, so the oceans look blue from out there. So, if you want to know why the sky is blue, why the planet looks blue from space, or why we kept the oceans that the asteroids so conveniently delivered, you need only one answer: the atmosphere.[13]

Filling the oceans is a pretty impressive job for the asteroids to have carried out. But there is a Big Question far more impressive than that: asteroids seem to have had a hand in the origin of life itself.

The Origin of Life

How life started is surely one of the most profound questions we can ask. For a long time it didn't even seem like a scientific question at all, but something that only philosophy or religion could address. How do you even start to give a scientific answer? But since the discovery of the double helix structure and genetic function of DNA in the early 1950s we have learned enough about how life works, about the extraordinary range of conditions where life can flourish, and also about the conditions that prevailed on the newly formed Earth that addressing the question of life's beginnings scientifically has become merely ambitious, not foolhardy.

Reflecting this progress, programs at universities such as the Origin of Life Initiative at Harvard and the Carl Sagan Institute at Cornell have sprung up. They have fascinating talks that I like to go to, but sometimes they get too biological for me to follow. These new venues are focal points of a new field we call astrobiology. New institutes are needed because a wide range of expertise in extremely different disciplines is needed to find a scientific answer to this big question. Biology, astronomy, chemistry, geology, and atmospheric science are just the obvious starting few. Combining this expertise is not easy. Those in the disciplines all speak different specialist languages, so they first have to learn one another's jargon just to communicate. That won't happen without bringing researchers from each field together in the same place. Asteroid mining needs a wide array of skills too, and similarly broad institutions are likely to be created to help it grow.

For initiating life, the chemistry of the element carbon is key. Carbon is what diamonds and soot are made from. Carbon can form so many different molecules that it has its own subject: organic chemistry. It got this name because all life on Earth is made up of carbon compounds. That's why "carbon-based life-forms" is a popular phrase in science fiction. Other suggestions, for example, "silicon-based life-forms," have turned out to be much less interesting. (Although our own carbon-based life might be in the process of creating silicon-based life within our computers, a field called artificial life, or alife.)[14]

Carbon is seven times more common than silicon in the universe, almost as common as oxygen. Being abundant gives carbon another advantage for being the basis of life. Many organic molecules contain oxygen and hydrogen, which are also very common elements.

When life began about some 3.8 billion years ago, the Earth was quite young, only about 700 million years old. Would organic molecules naturally form then? Why weren't the new oceans that the asteroids had provided just made of water and mineral dust? Why was the primordial soup not just a stone soup? It has been known for about 70 years that if you take the molecular ingredients found in the Earth's early atmosphere, which had no oxygen in it, and send sparks of electricity through them to simulate lightning, you can make complex organic molecules. Harold Urey and Stanley Miller did this in a famous experiment back in 1953. They made abundant amino acids, suggesting that the early oceans also had them. Variations of the experiment since have shown similar results. So maybe there is no problem.

But asteroids were bombarding that early Earth. They provide an alternative answer. If the water of the oceans came from the bombardment of the newly cooled Earth by asteroids, they would bring along whatever else was in them. That likely included huge amounts of organic molecules. Were they the seasoning of the stone soup of the early oceans, enabling life to develop? This idea was first suggested decades ago by Christophe Chyba and Carl Sagan.[15] It's still looking like a good bet.

Although space seems like a hostile place to us it's not really strange that organic molecules are found there. We already saw that water is common in space. The carbon we need to make organic compounds is, like hydrogen and oxygen, one of the most cosmically abundant elements. Carbon, like oxygen, is made in stars and dispersed in supernova explosions into the interstellar space of the Milky Way. Gravity then collects these elements into dense clouds out of which new stars, planets, and—crucially, it seems—asteroids form. In these dense clouds organic molecules form. This was a huge surprise when the first millimeter-band telescopes back in the 1970s began to find a rich organic chemistry in them. Even more complex molecules are found in the high-density, cool disks around young stars from which planets form.[16] Their environments in these clouds and disks are shielded from ultraviolet and X-ray radiation shining from the hot young star at the center. That allows these more delicate molecules to survive. Maybe, then, the ingredients for life came from space partly ready-made?

We know that there are many organic molecules in asteroids because carbonaceous meteorites can be rich in organic materials. "Rich" is no exaggeration. Glenn MacPherson is a senior scientist in the Division of Meteorites at the Smithsonian's National Museum of Natural History. When I visited his office a few years back, he told me that the famous meteorite that fell in Murchison, Australia, in 1969 is so rich in organics that when a new slice is freshly cut, it smells like asphalt or tar. That sounds promising! It tells us that many asteroids have rich collections of organic molecules within them. Laboratory analysis backs this up. Among the organic molecules found in carbonaceous meteorites are a variety of acids, not least many amino acids. Just 20 amino acids are used by life on Earth. Of these 20, only 12 have been found in meteorites to date. Are they really not present in asteroids, or did the other 8 get destroyed on the way to us? Does the primordial soup have to make these last 8, or have we just not looked carefully enough yet?

Proteins are formed from amino acids, and RNA and DNA are made up of proteins. It would be more persuasive that asteroid material seeded life if asteroids already contained proteins. Until recently, though, no proteins had been found in meteorites. Are there more complex organic molecules to be found in meteorites made from the polymerization of amino acids in space? (Polymerization is the formation of chains of molecules, either of the same kind or of alternating kinds.) Julie McGeoch of the Department of Molecular and Cellular Biology at Harvard and her colleagues have evidence that there are. McGeoch is a biochemist who came into the field of meteorite research in 2008 when working with one of the most ancient proteins on Earth, the rotor subunit of the enzyme ATP. (Technically, ancient proteins are "highly conserved.") ATP is a crucial molecule to life as it provides energy in cells. McGeoch decided to search for similar proteins in meteorites. Her searches have revealed long polymerized molecules and at least one protein, hemolithin.[17]

The advantage that McGeoch has over others in the field is that most meteoriticists are trained in geology, understandably enough. They are not generally used to dealing with delicate molecules. So for many years they have been extracting organic molecules from meteorites by boiling their samples for hours. This is not a good way to treat long molecules. As cooks know, heat changes the chemistry of organic materials such as meat and vegetables. Instead, McGeoch and her colleagues used the gentler techniques of biochemistry. They also took care to obtain their samples from deep inside the meteorites and studied them under clean room conditions to avoid contamination by Earthly molecules. The result was that they found molecules hundreds of atoms long. If this team is correct, then there is a lot more organic chemistry in asteroids than we had thought. That primordial soup may have been a lot tastier than anyone had guessed.

McGeoch's work is still controversial among meteoriticists. By going to a carbonaceous asteroid and taking a pristine sample we could find out definitively. Until recently no spacecraft had visited a carbonaceous asteroid. But two spacecraft are about to bring back samples. Japan's *Hayabusa2* picked up returned some rocks from the surface of Ryugu after a three-and-a-half-year voyage. NASA's *OSIRIS-REx* reached the asteroid Bennu in 2018 after a two-year voyage and will bring back as much as a kilogram of rocks from its surface. These spacecrafts' view of the chemistry in the early Solar System may well be a revelation.

So an early bombardment of the Earth by asteroids containing both water and organic molecules could have been responsible not only for filling the Earth's oceans but also for seeding the "primordial soup" with the ingredients for life. Asteroids may have accelerated, or even enabled, the formation of life. It's a great story. Going to the asteroids will tell us if it's correct.

The extreme possibility is that life wasn't just made easier to start on Earth thanks to asteroids providing the raw materials, but that life actually started somewhere else and was brought to Earth by meteorites. This is the idea of panspermia: that life spreads through space rather than starting on each individual world. This theory has a long history in modern science, going back to Jöns Jacob Berzelius—whom we will hear more about soon—in 1834, but had little currency until the discovery of those organic molecules in interstellar clouds.

Famed astrophysicist and cosmologist Fred Hoyle was a strong advocate for panspermia. Although he was widely laughed at by experts when he proposed his version of the idea, it is not wise to dismiss any of his theories out of hand. Hoyle was a very smart and creative scientist. His biggest undoubted triumph was to show, back in 1957, that the elements of the periodic table were made in stars. That's the origin of the idea that we are "stardust." He didn't prove this only in a general sense, but in extreme detail. Starting with almost pure hydrogen, he and his colleagues Margaret and Geoff Burbidge and Willy Fowler worked through the many nuclear reactions involved for every element and showed just why the elements are found in the proportions we actually see in stars and meteorites. This was Nobel-worthy work, but only Fowler got the prize because by the time he did, in 1983, Hoyle was considered too wacky. (Or so it is generally believed. The Nobel Prize committee does not explain itself.) Largely that was because he continued to push his "steady state" cosmology theory long after the discovery of the faint black body glow in the radio sky that was a prediction of the competing Big Bang model. Yet the mathematics Hoyle and his colleagues used turns out to be closely related to that used for the highly respected "inflationary" models.[18] Perhaps there is some deep connection between the theories.

Hoyle's wackiness score went up even more when he and his longtime collaborator Chandra Wickramasinghe proposed that life started on comets or, more generally, on the planetesimals beyond Uranus and Neptune, and was brought to Earth on meteoritic dust.[19] They were inspired by the 1970s discovery of organic molecules in space and by the discovery of amino acids in the Murchison meteorite in 1970.[20] They coined the term *prebiological* for these molecules, a term that has come into common use. Nevertheless, Hoyle's panspermia was too much for most scientists, and the idea disappeared from scientific discussions. Perhaps the samples coming back with *Hayabusa2* and *OSIRIS-REx* will have even greater surprises in store for us.

A different form of panspermia got a boost in 1996 when three scientists, David McKay, Everett Gibson, and Kathie Thomas-Keprta of NASA's Johnson Space Center in Houston, showed several lines of evidence that a Martian meteorite named ALH84001 contained tiny fossils of microbes.[21] (The cryptic-seeming name just means it was the first meteorite of 1984 found at the Allan Hills in Antarctica.) That life arrived from a planet— Mars—is a much less far-out idea than that life started on comets. The paper was well received at first, but all the evidence was immediately reexamined carefully by scientists around the world trying to look for flaws or alternative explanations. This is how science becomes reliable. Twenty years later most experts don't find the result convincing. Although the evidence was inconclusive, it did play a big part in restarting the NASA Mars exploration program. In fact, the whole field of "astrobiology" became respectable and flourished, not least because of the impetus given to it by ALH84001. Tests of panspermia, by putting known microorganisms into space for extended periods, is now a respectable research area. Given Hoyle's track record, we have to wonder if panspermia was a case of premature rejection. Samples from comets and asteroids, including interstellar ones like the first one, 'Oumuamua, and even interstellar meteors, will eventually answer the question.[22]

That leaves one last Big Question. This one takes us from life to civilization. How do the asteroids feature in that?

Origin of Ores

Without iron and other metals, our civilization would be technologically limited. We would not be able to use those metals easily if there weren't places on Earth where there were highly concentrated ores of them. Most of the 80 or so useful elements need to be concentrated somehow for us to have found them or to mine them. Then we can use them to build our myriad tools of enormous power. A metal plow opens up far more land for agriculture than a wooden one, for example. Almost all the industrially useful elements in the Earth's crust, with the sole exceptions of iron and aluminum, make up just 1/10 of 1 percent of the crust. And yet there should be even less. Where the precious metals in the Earth's crust come from is a long-standing mystery. It turns out that asteroids have a role in this too.

As with Olbers' paradox, asking *Why?* about something we all know to be the case can be deeply revealing. The existence of rich ores of heavy metals in the Earth's crust turns out to be surprisingly interesting. Why aren't precious metals scattered smoothly around the Earth rather than being bunched up in a few special places? Surely plate tectonics and volcanoes and such heat up rock and allow the gold and other rare metals to distill out in a few places. Well, yes, they do; but once you start thinking about it you realize that the first question you should ask is "Why are there *any* metals that dissolve in liquid iron—the siderophiles—in the Earth's crust at all?" After all, didn't we just learn that they all sank to the center of the Earth along with the iron back when the Earth was molten?

The most promising idea is that the heavy metals that the Earth was born with really did sink to the iron core, the way that physics says they should. But then asteroids brought a fresh supply here sometime after the Earth had cooled enough for the crust, the part we live on, to form. The rain of asteroids that did this came even before the Late Heavy Bombardment brought on by Jupiter's wanderings. This is called the Late Veneer and is the tail end of the first growth of the planets. Despite being called a veneer, this rain of rocks brought much more mass to the Earth than the Late Heavy Bombardment.

How could you prove this idea was right? Or show that it was wrong, for that matter? The answer is to use nuclear isotopes. Each type of atom, each "element," has a particular number of protons in its nucleus, from 1 to over 100. The attraction of the positive charge of the protons keeps the negatively charged electrons bound to the atom. Atomic nuclei also contain about the same number of neutrons as protons. Neutrons have no electric charge; they are neutral—hence the name. Each element can have a slightly variable number of neutrons. Any of these variants is called an isotope. For example, hydrogen normally has one proton and no neutrons. But hydrogen has a form with a neutron (but still one proton—that is what keeps it hydrogen). This variation on hydrogen is called deuterium, whose chemical symbol is D. Water made of D_2O instead of H_2O is called heavy water.

In 2011 a small team from Bristol and Oxford Universities in the United Kingdom, Matthias Willbold, Tim Elliott, and Stephen Moorbath, used isotopes to test the idea that many of the Earth's precious metals—the siderophiles—come from asteroids.[23] To do so they analyzed the tungsten in some of the oldest rocks to have survived unchanged to the present. These are 3,800-million-year-old rocks from a small outcrop at Isua in Greenland.[24] That's an interesting age, coming about 700 million years after the Earth formed. By this time the Moon had already been ripped out of the Earth by a giant impact and things had settled down again, letting the crust form. The crust is the thin layer of the continents and the ocean bottoms. It makes up only about 1 percent of the Earth's radius. But it is a time before the Late Heavy Bombardment that the Nice model tries to explain, and before the Late Veneer, so the rocks would not be contaminated by either event. The team looked at two tungsten isotopes, ^{182}W and ^{184}W, in these ancient rocks. (W is the chemical symbol for tungsten, from the German *wolfram*.) If fresh tungsten was supplied by the Late Veneer, then the Isua rocks and later rocks will have different ratios.

That's exactly what the team found. The researchers concluded that tungsten in younger rocks likely came from asteroids. More recent work with the siderophile element ruthenium shows the same pattern.[25] How that tungsten got into the crust is a little more complicated. The late bombardment of asteroids, rich in metals, peppered the crust like buckshot. They went right through to the upper mantle, enriching it with tungsten, but also with gold, platinum, and the other useful elements. It took normal geological processes to concentrate some of these heavy elements into the veins of ore we mine today.

If the asteroids are as important to these Big Questions as they seem to be, then perhaps we have to thank asteroids, and the migrations of Jupiter, for our very existence. There are many planets circling other stars that lie in their "habitable zones" where water could be liquid, but maybe only a small fraction had asteroid bombardments thanks to wandering gas giant planets. If that's the case then there may be many worlds in other solar systems that could have supported life but remained sterile because they were never blessed with oceans and the organic matter from which life could spring, or with the iron and other heavy elements from which to build technology had life emerged. All for want of asteroids hitting them.

But enough of love. The asteroids also have a dark side. Let's move on to fear.

3

———

Fear

In the movies our small planet is constantly being threatened by aliens and asteroids. Should we worry?

During a lunch discussion about aliens one day in 1950, some Los Alamos physicists were having fun trying to estimate how many worlds in the Galaxy were inhabited by aliens. They decided there must be lots of them. Then one of them, the great Enrico Fermi, stumped the rest when he asked simply, "Where are they?" In other words, if they are so common, why haven't we seen them already? This turned out to be another of those simple but profound questions, like Olbers' paradox. Fermi may or may not have realized this at the time. By now there's a huge—but totally inconclusive—discussion about this "Fermi Paradox." One idea is that those aliens are out there; they found us when we were just promising apes and left a trip wire to alert them to when we grew up. That's the reason there's the monolith on the Moon in the movie *2001: A Space Odyssey*. Or, as Charles Stross suggests in his novel *Accelerando*, maybe advanced civilizations get so addicted to high-bandwidth communications that they stop venturing far from their home star. They are social media shut-ins. My favorite answer is Terry Bisson's in his micro-story "Meat."[1] Some patrolling aliens do find us but are disgusted to find that we are made of meat, "meat all the way through." We are so repulsive, in fact, that they fake their log to say no intelligent life was found because, well, ick! For whatever reason, aliens seem to be rare. And anyway, we can't do much about aliens.

That leaves the asteroids. Should we fear the asteroids? When I talked about love I played a trick on you, substituting the love of knowledge for the romantic love you probably expected. This time though, when I say fear, I mean fear—for real. Could a killer asteroid slam into our planet and destroy us, the way one destroyed the dinosaurs 65 million years ago? Bluntly, yes. Large asteroids have hit the Earth since life evolved, with catastrophic effects. Eventually, another will do so.

Although asteroids no longer pound the Earth, raining down oceans of water, Earth is still in a shooting gallery. Asteroids were called the "vermin of the skies" by famed astronomer Walter Baade.[2] He meant that they get in the way of "real" astronomy—stars and galaxies. But since he said that, over 50 years ago, we have come to realize that asteroids are vermin also because far too many of them come scuttling by uncomfortably close to our small blue planet. These are called "potentially hazardous objects" (PHOs). (They are called "objects" rather than "asteroids" because there are some comets hiding among them.) They are a subset of the near-Earth asteroids. One of them really could be our Armageddon.

Having flying mountains and boulders coming close to the Earth in the form of near-Earth asteroids is dangerous. Some of those asteroids get very close indeed; some in fact do hit our planet. It's smart, then, to keep a watch list of any asteroid that might hit us. A major focus of the International Astronomical Union's (IAU) Minor Planet Center is to identify these dangerous asteroids that might hit the Earth sometime in the future. The ones on the organization's watch list must have an orbit that could bring them close to the Earth's orbit, within about 20 times the distance of the Moon. Officially, to qualify as a potentially hazardous object it must also be large enough to reach the Earth's surface (as small asteroids disintegrate in the Earth's atmosphere) and be highly damaging. This is taken to mean a diameter of 140 meters or larger. That's about the size of a football stadium. This size criterion is a bit arbitrary. An asteroid of just 20 meters diameter, the size of a big house, can reach the surface too. One of those could wreak a lot of havoc on a local scale.

There are over 1,000 potentially hazardous objects known at present, making up about a quarter of all the known near-Earth asteroids larger than 140 meters in diameter. As smaller asteroids have to be quite close to the Earth to be bright enough to detect, a larger fraction of them, about

40 percent, come close enough that they would qualify as PHOs, but for their relatively diminutive size. None of them yet have a scarily high chance of hitting the Earth in the next century or so. But we've only found a small minority of them so far.

We do know about some of the impacts that have happened in the past. Starting with the most recent, let's look at some of them.

Chelyabinsk

What will people do when an asteroid hits the Earth? Obviously they would run screaming from the fireball streaking across the sky threatening their total destruction. That's what everyone does in the movies. You'd run for your life, wouldn't you? Well, now we know what people *really* do when faced with "death from the skies."[3] We do just the opposite! The day after Valentine's Day in February 2013, an explosion equal to half a million tons of high-explosive TNT rocked the skies over the Siberian city of Chelyabinsk.[4] That's the energy of 50 Hiroshima atom bombs. Chelyabinsk is one of the largest cities in Russia, with over one million inhabitants. Dozens of dash-cams, which are common in Russian cars to record encounters with the police, recorded the huge streak across the sky. None of the dash-cam videos showed anyone even stopping their cars. At least one person turned in the direction of the trail across the sky, presumably to follow it. (That individual would have been too slow only by 60,000 kilometers per hour!) Lots more people indoors went to their windows to see what the bright flash was. The result of the experiment? Love, or at least curiosity, beats fear.[5]

No one was killed in the Chelyabinsk event, but 1,600 people were injured by the blast, mostly by glass shattering when the sonic boom arrived at their windows a couple of minutes after the flash of light. Some people close to the path of the meteor were temporarily blinded by the flash, while a few got sunburned.[6] If the fireball had been a little lower that "sunburn" could have turned into major burn injuries.

The energy released into the atmosphere by the Chelyabinsk asteroid-turned-meteorite is really well measured. The fireball created huge low-frequency sound waves—infrasound—that traveled all around the Earth several times. This infrasound signal was picked up by a worldwide network of sensitive detectors set up to monitor nuclear weapons tests. How big was this rock that was a potential city destroyer? Just 17 meters across, the size of a large house.

The track of the asteroid across Siberia and the speed it was traveling could also be measured accurately. Traffic cameras monitoring the main square in Chelyabinsk captured the shadows of the lampposts in the square created by the bright glow of the fireball. As the fireball went by, the shadows swept around. Using just the distance between lampposts and a little simple geometry, it didn't take a scientist to work out how far away the fireball was and so how fast it was moving. In fact a blogger, Stefan Geens, did exactly that the day after the event using Google Earth to get the distance between the lampposts.[7] Other videos backed up his calculations. A few months later Jorge Zuluaga and Ignacio Ferrin, two scientists in Medellín, Colombia, refined the numbers and found that Geens's first estimate was close to the mark.[8] The speed they found, 19 kilometers per second, is too fast for a ballistic missile and too slow to be a comet, but just right for a near-Earth asteroid.

If the blast was as huge as a nuclear bomb, why wasn't the whole city of Chelyabinsk flattened? That the track was so high and almost horizontal is what prevented a much larger disaster. The long delay before the sonic boom arrived tells us that the fireball was high in the sky. The speed of sound is only 1,000 kilometers per hour, so the fireball must have been brightest when it was about 30 kilometers high. The meteorites from the Chelyabinsk asteroid are stony and so quite easily broken up. That's why it disintegrated relatively high up in the atmosphere instead of lower down, where the heat and shock wave would have caused more damage. The asteroid was just skimming through our atmosphere. If it had been heading straight down it would have deposited almost all its energy in a short time and in a smallish spot on the ground. That would have been much more like an atomic bomb exploding, and similarly devastating, though without the radioactive fallout. There would have been no time then to go to the window or turn to follow the fireball.

How often will one of these mansion-sized rocks hit the Earth? Once every 100 years, according to the best estimate of Alan W. Harris. But Chelyabinsk has led to a recalibration that postulates they are some 10 times more common.[9] We can expect one of these hits every decade or two. Just six years after Chelyabinsk another fireball that had almost as much energy exploded over the Kamchatka Peninsula in Russia, near to the Bering Strait that separates Russia from Alaska.[10] In earlier decades it would have been missed, but there are now enough instruments monitoring the globe that it was found. Are we underestimating the rate then? One pair of impacts in a few years doesn't tell us because these events are random. Paradoxically, random events come in groups. After all, if they came regularly then they wouldn't be random. We would need to find a way of determining all the impacts several centuries back before we could be sure of the rate.

As a result of the Chelyabinsk near disaster, Lindley Johnson, NASA's planetary defense officer, has revised the asteroid size that NASA wants to find and track from 100 meters to 50 meters.[11] That is still bigger than the Chelyabinsk rock, but choosing a 20-meter threshold right now would make little sense, as we haven't come up with a way to find most of them. There are some 10 million near-Earth asteroids that are at least that big.[12] Of that multitude we have seen only about 1,500, and many of them were lost again soon after they were found. On the other hand, perhaps declaring a goal of finding all asteroids down to a size of 20 meters would set astronomers to thinking creatively about just how we might do that.

True, most of the Chelyabinsk-sized asteroids, like all others, will land in the oceans or somewhere that causes no harm to people, but it is still sobering that they may not be that rare after all. Is it time to get worried?

Tunguska

It's good to examine the historical record to see if we can get a handle on how often Chelyabinsk-scale impacts happen.

There was at least one other largish asteroid that hit the Earth in modern times. The famous 1908 "Tunguska event" was more like a bomb explosion. Tunguska is also in unlucky Siberia, but fortunately it lies in a much more isolated spot than Chelyabinsk. Had the asteroid arrived slightly earlier or later it could have landed on a much more densely populated location in western Europe or China, where it would likely have done a lot more damage to people and property. The energy released by the Tunguska event was at least 20 times bigger than Chelyabinsk, probably around 10 to 20 megatons, or 1,000 to 2,000 times the energy of the Hiroshima bomb.[13]

In 1908 there was no infrasound network to measure the energy so we have to work it out from other clues. Finding those clues took a while. Tunguska was not investigated at once, and then World War I and the Russian Revolution distracted attention from such obscurities. Finally an expedition led by Leonid Kulik visited the site nearly 20 years after the event. Kulik's team found that the area of Siberian forest knocked flat by Tunguska is some 15 kilometers around the point below the impact. That's an area of 800-plus square kilometers, the size of a major city, such as the Washington, DC, metro area. A Tunguska-class impact is definitely a city killer.

Early in 2018 I dropped in on the NASA Ames Research Center in Silicon Valley for a one-day discussion of the latest scientific results on Tunguska. It was a small room, and the discussion attracted about 30 scientists, quite a bit more than the organizers had expected, so it was rather crowded. Still, the talks were fascinating. The downed trees formed a pattern radiating out from the point below the impact. Knocking down that much healthy forest would have taken 5 megatons. Mark Boslough, of the Sandia National Laboratory in New Mexico, has suggested that the forest at the impact point was dying and so was more easily felled.[14] The latest estimates put the Tunguska asteroid at a diameter of 50–70 meters, or half a football field. That's only a bit more than twice the size of the Chelyabinsk asteroid. You may be surprised then that it packed a 10 times bigger punch. Actually, that's about what we'd expect. The energy of an asteroid's motion is proportional to its mass. Being twice the diameter means having 8 times the volume and so mass, if it is made of the same type of rock. So that works out right. It's always good to check though!

Wabar Craters

There are other quite young impact sites on Earth. A listing of 190 confirmed impact structures with their estimated ages is maintained in the Earth Impact Database hosted by the Planetary and Space Science Centre at the University of New Brunswick, Canada.[15] It lists three more impacts that are less than 200 years old. They were formed by asteroids listed as being some 13, 20, and 110 meters across. The small one fell in Carancas, Peru. The midsize one fell in Russia (again!), and the large one fell in Saudi Arabia's fearsome Empty Quarter. These are the Wabar craters. They were first reported by St Jean Philby in 1933, and were then pretty much ignored until Gene Shoemaker and his colleagues undertook expeditions there in 1994.[16] They were able to date Wabar to between 1545 and 1859. This is just about consistent with Philby's account of stories of a fireball that was seen passing over Riyadh in the late nineteenth century, although that date is uncertain too. The direction of the fireball does agree with Shoemaker's investigation. Another suggested date of 1704 agrees better though. It comes from two poems that relate a fire in the sky on the same Saturday night in that year.[17] Wabar was quite like Chelyabinsk. It came streaking in at a low angle at about the same speed as Chelyabinsk. The difference is that it was an iron asteroid weighing 3,500 tons, and so made larger craters than Chelyabinsk.

So from 1704 to 2013 we know of five events involving Chelyabinsk-sized impacts or larger.[18] That's one every sixty or so years. That seems to be a little slower rate than the rate predicted by Alan W. Harris. Our lists must be incomplete though. For one thing, the cataloged craters are conspicuously concentrated in just a few parts of the globe: Canada, Australia, and eastern Europe (see figure 4). This bunching up must be at least partly a result of where they have been looked for most carefully, and also of where they are most easily preserved. Evidence of craters can be destroyed quite quickly in some locations. The Wabar craters have rapidly filled with sand. When Philby was there the main crater was 13 meters deep; just 60 years later the Shoemaker expedition found it to be only 2 meters deep. It might well vanish from sight in a few more decades. That means there could be other, yet unrecognized craters from the past few hundred years still waiting to be found, and some that are no longer findable. So once per decade or two seems like a good rate to work with.

The number we're aware of in the twentieth century does suggest that they are common enough to be scary. How do we know we are not about to be hit by something big enough to wipe out a large city and its millions of inhabitants? We don't. And there's worse to come.

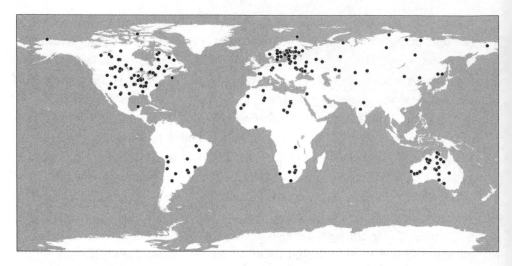

Figure 4: World map of the 190 confirmed impact craters. Note that they are not uniformly distributed, as one would expect. There are probably more that have eroded away or escaped detection.

The Barringer Crater

Neither the Chelyabinsk nor the Tunguska event led to a large piece of an asteroid hitting the ground and forming a crater. But our planet is not always that lucky. Once we go back to the older craters we find some much nastier impacts.

The Barringer Crater is one of the most famous and best-preserved asteroid impact sites.[19] This 1200-meter-diameter crater in Arizona is also known as "Meteor Crater." It was formed some 50,000 years ago when a roughly 30-meter-diameter asteroid slammed into northern Arizona, creating a hole deeper than the Pyramids are high. For a few seconds the crater was twice as deep, but it was immediately filled in again by rock falling back down. The energy it released is thought to be somewhere between 20 and 40 megatons, enough to kill half the large animals within 10 kilometers. Forty megatons is "only" a Tunguska-class event. So why did it leave an impressive crater when Tunguska did not? Barringer shows us what happens when an asteroid made of solid metal hits the ground.

Because asteroids are made from a wide range of different materials the impact they make can vary a lot. The difference between a 40-meter-diameter rock and a 40-meter-diameter block of solid iron is quite dramatic. The more crumbly rock tends to explode in the atmosphere, like the Chelyabinsk and Tunguska events. That's because, as it travels far faster than sound, the air cannot move out of its way and huge pressures build up on the rock, easily separating the loosely bound rock and rubble. All those small pieces burn up quicker than a single block. If the iron asteroid is a single piece, not a pile of fragments, then it can reach the ground while still traveling very fast. That may or may not be a fair assumption.

The Barringer Crater was made by an iron asteroid. From its diameter of about 1 kilometer it's possible in principle to work back to how much energy was released by the impact and so how large the incoming asteroid had to be. But the answer depends on how fast it was moving and what angle it came in at, neither of which are all that well known for the Barringer event. Even if we knew these, the calculations are tricky and require some serious computing. The best estimates are that it was somewhere between 10 and 50 meters across. For now we can't do better than this rather wide range.

The Barringer Crater is unusually well preserved and so has been studied in great detail. Its age is really well known: 49,000 years, with an uncertainty of just 3,000 years. How can anyone be so sure? Again, scientists use a link between the Earth and the cosmos. There are a number of places far out in space where atoms, protons, and electrons are accelerated up to speeds just a hair shy of the speed of light. They are called cosmic rays. These cosmic accelerators easily beat the highest energy particle accelerator on Earth, the Large Hadron Collider at CERN (the European Organization for Nuclear Research) in Switzerland. Cosmic rays are used to date rocks.

These fast-moving particles come from supernovae that are quite nearby, just a few thousand light years away in our Galaxy. Supernovae are the exploding stars that disperse elements into the Galaxy, and as the shock wave from their explosion spreads out into the surrounding tenuous gas between the stars (the interstellar medium, we call it), they are accelerated into cosmic rays. How they do this was first worked out by Enrico Fermi, whom we met earlier when we talked about aliens. He created his theory in the 1930s, but the strongest proof of his idea did not come until 70 years later, and fully 100 years after the discovery of cosmic rays, thanks to observations with the Chandra X-ray Observatory.[20] I'm really happy that Chandra data solved this stubborn problem, as I spent many years doing my bit to help that observatory operate successfully. These fast-moving particles pervade the whole of our Milky Way Galaxy, so they are called galactic cosmic rays to avoid confusion with other fast particles streaming out from the Sun.

Galactic cosmic rays give us a clock for determining how long it's been since the rock froze out of the molten state. They are constantly raining down on our atmosphere from all directions. When they collide with an atom of the atmosphere, they produce a "shower" of other particles all moving in much the same direction as the initial cosmic ray. Many of the particles in the shower reach the ground with enough energy to disrupt rock crystals, creating "defects." The more defects, the longer the rock has been exposed to cosmic rays. The rocks melted by the Barringer event started out with no defects then but have accumulated defects ever since. To read this clock the rock crystal is heated. This heals the defect and, in doing so, it emits light. This is called thermoluminescence. The more light is emitted, the longer a rock has been solid. With a lot of care this can yield a very precise measurement of just how long it is since a rock formed. In the case of Barringer that is 49,000 years. The same technique was used to date the Wabar craters at less than 400 years old.[21]

This crater was named after Daniel Barringer, who originally made his fortune as a silver miner. Barringer was the first person to realize that the crater had been created by a meteor. In 1905 (the same year that Einstein published his theory of special relativity), after careful investigation of the site, Barringer published his claim that the crater had been formed by a large meteor impact. That was a novel idea and went against the explanation by an eminent Smithsonian geologist of the time that the crater was formed by a volcano. (Scientists from my institution are not *always* right.) Barringer's arguments were strong and were based on better site work than the eminent geologist had available. Nevertheless, it was more than half a century before his explanation was accepted. Partly this may have been the preference of academics to believe one of their own rather than an outsider, and a miner at that.

Their reluctance to accept the idea of an asteroid impact may have been strong because geologists had by then been arguing for decades that "deep time"—measured in hundreds of millions of years—was needed to form rocks from silty lake bottoms, raise mountain chains, or carve canyons. They had turned this into a principle: that only processes we see happening today happened in the past. They called this the uniformitarian principle, first named by James Hutton, a Scottish naturalist, in 1785. (Meteorites were first proposed to be rocks falling from the sky just a few years before.) In astronomy we call the same idea the Copernican principle. The claim is that we are nothing special; the Earth is not the center of the universe. It sounds so humble; but really it is extreme hubris—the whole of the universe behaves just like an apple falling from a tree? What an arrogant assumption! And yet it works. We astronomers routinely apply laws of physics that were established in the laboratory to vast galaxies far away, with excellent results.

The alternative to uniformitarianism is catastrophism, invented in the same exciting time for science near the end of the eighteenth century. This idea posits that rare, dramatic, and brief events can have important, even global, influence. The French naturalist Georges Cuvier was the most influential proponent of catastrophism. In 1796—only five years before the first asteroid was found—he realized that mammoths and mastodons were not the same species as either African or Indian elephants, so they must be extinct species. The idea of extinction was new, and quite a shock. Cuvier went on to find that major extinction events had happened several times in the history of the Earth, an idea he laid out in the year of Napoleon's defeat in Russia, 1812. (By then there were four asteroids known.) So for him catastrophism was a reasonable position. But it was a poor fit with the uniformitarian application of known processes on a grand scale that was working so well for physics and geology.

Now we understand that both ideas are right—only natural processes are needed to create what we see in both geology and astronomy—and most act exceedingly slowly. But occasionally a single instant can wreak major changes through a rare violent event. A supernova is one dramatic example. The favored origin story of the Moon is also brief and violent— that it was torn out of the early Earth during an impact by another Mars-sized planet. A big asteroid hitting the Earth is another. It is rare, but when it happens it's quick, and it's a big deal.

The uniformitarian principle breaks down because human lifetimes are so short compared with geological and astronomical timescales—at least a million times shorter. So when we deal with events that are rare on a human timescale, yet common enough on geological timescales, we often miss them.

It took the invention of atomic bombs to convince scientists that the Barringer Crater was made by an asteroid impact. The craters that bomb tests made in the 1950s provided the young Eugene Shoemaker with good analogs to the Barringer Crater. Times had changed, and in 1963, when Shoemaker published his PhD work on the geology of the Barringer Crater, he nailed the case for an impact.[22] Perhaps his insider status as someone with a Princeton PhD helped, but his evidence for an impact was strong. With Edward C. T. Chao, he found a telltale mineral, coesite, near the crater. This is a form of quartz that forms only in powerful shocks. He also showed that as you go up the layers of rock in the crater wall you find that midway through the sequence repeats in reverse order. The rock was folded over by the force of the impact. His work was accepted by other geologists, and his discovery led him to found the new field of astrogeology.

Gene Shoemaker was the first person to bring fear into asteroid research. After spending nearly a decade preparing for and taking part in the Apollo program, he moved to the California Institute of Technology (Caltech), where he started the search for asteroids in Earth-crossing orbits that might one day collide with Earth. We now call these the near-Earth asteroids. Near-Earth asteroids turn out to be really important not only for fear, but also for love and, in the not too distant future, for greed.

Barringer himself didn't wait for scientific acceptance. President Teddy Roosevelt had already granted him mineral rights to the land around the crater back in 1903, and he set about trying to mine the large iron meteorite he expected must be there. He estimated it was worth $1 billion in 1905 dollars. That's at least $25 billion today, so you can understand his motivation. Alas, he spent 27 years searching in vain. The asteroid, we now realize, had been destroyed in the impact. Barringer's descendants still own the crater and the land around it. They have found tourism to be a much better business than mining. Their proceeds help support an annual grant for meteor crater investigations. We'll come back to both tourism and mining when we get to greed.

Fear is the subject now. The air blast from the Barringer event did a lot of damage. The energy it released was similar to the largest nuclear bomb tests from the 1950s and 1960s. The largest U.S. bomb was Castle Bravo at 15 megatons. It was exploded over Bikini Atoll on 1 March 1954. The Soviets exploded a much larger bomb, Tsar Bomba (Emperor of the Bombs), at 50 megatons on 30 October 1961. The Barringer impact was somewhere in the same range. Using nuclear bomb tests as a guide, David Kring of the Lunar and Planetary Institute in Houston has estimated that at Barringer trees were flattened by hurricane-force winds up to 30 kilometers away from the impact point, an area of nearly 3,000 square kilometers.[23]

Barringer's meteorite was on the small end of the impactors we have to worry about. Even an iron asteroid would not be likely to reach the ground if it were smaller. Iron asteroids are only a minority of asteroids, about 1 in 20. But we get one or two impacts of this size every century, so a Barringer-like impact should happen every millennium or thereabouts. That's somewhat reassuring. Barringer, though, is small potatoes compared with the extinction-creating impacts, the most famous of which happened 65 million years ago.

Chicxulub: Killing the Dinosaurs

The dinosaurs, as everyone now knows, were wiped out by an asteroid impact some 65 million years ago. A lot of evidence gets us to this startling conclusion. The culprit everyone points to is the Chicxulub impact. It hit just off the coast of what is now the Yucatán Peninsula of Mexico, near where the town of Chicxulub is today. So we know that there was an impact right about when the dinosaurs disappeared, and it must have been a big one, judging from the size of the underwater crater it left behind. The asteroid that made the Chicxulub crater had to have been some 10 kilometers across. That's 500 times the diameter of the Chelyabinsk asteroid and about 125 *million* times more energy—over 60 billion tons of TNT. It's 1,000 times more energy than the biggest nuclear bomb ever exploded. You would expect that to have pretty nasty consequences! That it might kill off much of life on Earth does not seem too far-fetched.

The first evidence that an asteroid did kill off the dinosaurs was the iridium layer—a centimeter-thick layer of rock found at the junction between the Cretaceous and the following periods, with the newer rocks on top. (Once the later period was called the Tertiary, giving rise to the term K/T boundary for this layer. Why K? It comes from the German equivalent of Cretaceous, "Kreide." Both names mean "chalk." But Tertiary is no longer an officially recognized name. The new term for the next period is the Paleogene. So the boundary is now called the K/Pg boundary.) It's called the iridium layer for the good reason that the element iridium is more than 10 times as common in this layer as it is in the Earth's crust, where it is very rare. Even so, only 6 parts per billion of the iridium layer is actually iridium. The striking thing about the iridium layer is that below it there are dinosaur fossils; above it, there are none. That's circumstantial evidence that the iridium layer had something to do with their disappearance.

How do you get a layer rich in iridium? Well, iridium is up to 100 times more common in asteroids than on Earth. So ascribing this excess of iridium to coming from an asteroid is plausible. The iridium layer is found worldwide, so it would have to have been a big asteroid. (This is an example of an asteroid bringing heavy metals to Earth as probably happened in the Late Veneer and the Late Heavy Bombardment.) Vaporized iridium-rich asteroid rock was spread over the surface of the whole Earth by the impact, creating the iridium layer. The iridium layer also contains a surprising amount of ash, suggesting a whole lot of burning forests went along with the arrival of all this iridium. That pollution in the atmosphere could have led to a years-long winter.

An odd thing about the K/Pg extinction event is that it was selective. Of the dinosaurs, only avian ones survived. We call them birds. Apparently, it was only the avians *not* living in trees that survived. Daniel Field, a paleontologist at Cambridge University, and his colleagues believe that the fossil record shows that the forest canopies collapsed after the impact.[24] So perching birds went extinct because they had no more trees to live in. Ground-dwelling avian dinosaurs survived, along with reptiles, insects, and the small mammals from which we eventually evolved.

But if the world was in darkness for years and forest canopies collapsed due to fire, shock waves, or years without sunlight, how did any land animals survive at all? A study of over 3,000 fossilized birds' teeth by Derek Larson, a paleontologist at the Philip J. Currie Dinosaur Museum in Alberta, and colleagues at the University of Toronto and elsewhere shows that the surviving birds all had beaks, like modern birds, while the toothed birds died out.[25] They suggest that the survivors may have been the ones that fed on seeds. Seeds are hardy and could have supplied food when all greenery was destroyed, but only for those who could eat them. It's fair to say that this is still not settled science.[26]

Are we sure that a killer rock from space is the right answer? There was a whole host of other ideas for what killed off the dinosaurs before the iridium layer was found.[27] Almost all of them are ruled out because they don't fit the facts. There is still one serious competitor: vulcanism and plate tectonics. Which is right? The volcanoes versus asteroids dispute has been a fierce one. A 2018 article in the *Atlantic* by Bianca Bosker called it "the nastiest feud in science."[28] It is a replay of the uniformitarians versus catastrophists disputes, with the volcanoes being the uniformitarian position.

Volcanoes are promising because iridium is also relatively common in the Earth's mantle. The lava from volcanoes is rock from the mantle, so it too contains more iridium, which could explain the iridium layer. Forests would be ignited by the molten lava, explaining the ash. Perhaps Earth is quite capable of creating a mass extinction all by itself, without extraterrestrial help? There were in fact enormous volcanic eruptions at about the right time. They formed the Deccan Traps in India, covering 500,000 square kilometers to depths of up to 2 kilometers. That's a lot of lava. The volcano case has been championed by Gerta Keller, a paleontology professor at Princeton University.[29] A key argument she uses to favor the volcano side is that life was in decline for several hundred thousand years before the Chicxulub event. This claim is based on the numbers of single-cell marine creatures called foraminifera, which seem to decline in numbers over that length of time. An asteroid couldn't cause that. But a giant series of volcanic eruptions, namely the Deccan Traps, could. This dispute has raged for 30 years, so far. A lot of other specialists disagree with Keller's measurements of foraminifera numbers.[30] Eventually we will know, as paleontologists continue to do the painstaking work of pinning down exactly when the Deccan Traps eruptions took place.

Not being 100 percent sure whether to blame killer rocks from space or volcanoes is totally normal in science. Having different theories and testing them against their predictions is how science works. Living with uncertainty is the fate of all scientists. Finding new information that doesn't fit any of the prevailing theories in an obvious way is also normal. In fact, this is the fun part of science, trying to solve a problem that has stymied others. There is no doubt, though, that a giant impact happened at the time of the Chicxulub event. It left a crater over 160 kilometers across, 20 times larger than Barringer. Faint traces of the crater are visible even underwater, and shattered rock fragments that are found only near asteroid impacts ring the area. That can't have been healthy for the remaining dinosaurs.

Now it's time to really *fear*. There are other killer asteroids out there. Even if we are wrong and the dinosaurs were actually killed by volcanoes, not knowing which theory is correct shouldn't stop us acting on the asteroids. A multi-kilometer-sized rock *will* hit the Earth really fast some day. That would not be a good day for humanity.

There are about 1,000 potential dinosaur (or rather human) killers out there, the ones bigger than a kilometer. Even before a 2005 congressional mandate, the George E. Brown, Jr. Near-Earth Object Survey Act, NASA had sponsored searches for asteroids, beginning with Spaceguard in 1998. These have found 90 percent of the really big ones, and none of these is any threat for now.[31] But it takes only one. Finding the last 100 will not be easy. Some could be lurking behind the Sun, in orbits that only slowly bring them toward the Earth. We may see them only when it's too late. Meanwhile, the Tunguska-sized, city-killing asteroids are 100 to 1,000 times more common than the dinosaur-killing kind. We'd like to find all of those too.

What Are the Odds?

Just how scared should we be? How big is the threat, realistically? To answer that we have to know how many big asteroids are out there that could someday collide with our home planet. How do we find these dangerous asteroids? Many potentially hazardous asteroids have been found in the last decade—but that's because we've only really been looking hard for them in that time. Our searches are still highly incomplete. We miss almost all the small ones and any that approach us from the daylight side of the sky, as the Chelyabinsk asteroid did. While the probability of one

hitting in the next century is *thought* to be low, how much of that assessment is due to our missing key events? Any that impacted the oceans far
from the shipping lanes, or in other remote regions like the Sahara Desert,
might not have been noticed until the past few decades, when worldwide
monitoring data (collected to monitor the 1963 Partial Nuclear Test Ban
Treaty) was made available. So we have very little direct information about
anything that happens less often than once in a half century.

We have to count the rocks to estimate the size of the threat. Although
we haven't found all the asteroids, we can work out how many must be
out there. The standard astronomy approach is to carefully record how
much of the sky you have surveyed, how often, and how close an asteroid
of a given size and albedo could be yet still be undetected in your survey.
If you can see something that is 10 times fainter, then you expect to find
more objects, but if you look at only half the sky, you expect to find only
half the number of objects. Then you can work out how many asteroids
there must be for you to see the smaller sample of them that you did find.
This is a good method in principle, but it's quite complex to do this in
practice, not least because near-Earth objects move quite fast, and they
get brighter and fainter in just days. Also, near-Earth asteroids have been
found by a variety of surveys, and not all of the surveyors kept careful
records of where they looked. The two methods agree fairly well, though.
There are about 20,000 near-Earth objects at least as big as 100 meters
across, and only about 1,000 as big as a kilometer across.

Perhaps a simpler way to think about this is to compare it to collecting
baseball cards. If you get a duplicate card, it not only gives you something
to swap, it also gives you an idea of how many different types of cards there
are out there, even though you haven't seen every one of them. If every
card you get is a duplicate, then there's only one type out there. If you
never get a duplicate, then there must be many, many different ones.
Counting carefully and applying some simple statistics, you can make this
estimate more precise. Alan Harris has used this approach.[32] Luckily, it
turns out that there aren't many asteroids in the dinosaur-killer size range.

For the smaller ones, the simplest method is just to count how many
actually hit the Earth, or the Moon, each year. For anything as big as a
house, like Chelyabinsk, zero is the answer in most years. But there are
many more asteroids just 1 or 2 meters across that do hit us each year. We

know this from studying the infrasound—very low-frequency sound waves—that they generate on hitting the atmosphere, as happened at Chelyabinsk. Peter Brown and his colleagues at the University of Western Ontario measure them regularly.[33] A similar direct counting method is to watch the Moon looking for small flashes of light that can be seen whenever a small asteroid hits the nearside. These two complementary approaches imply that altogether there are about 20 million near-Earth asteroids bigger than 2 meters across. That's quite a lot.

It's one thing to work out how often we can get clobbered by a giant rock from space. But how do you decide how hard to work, and how much money to spend, on preventing that from happening? The approach used so far is to think like an insurance company. A first step is to work out how bad an impact would be from a given size of asteroid. Determining how many people would be likely to die from any given impactor seems like a good place to start, though property damage would have to be factored in too for a sound insurance plan. Then we need to know how often each impactor is going to happen, on average. That's where the surveys for potentially hazardous asteroids come in. Alan Harris uses these numbers to estimate how many people per year, on average, die from impacts. By doing this for each size of asteroid, he can compare relatively small Tunguska-like impacts from true Chicxulub-level dinosaur killers. This is sensible because it tells you how much you should charge for an insurance premium, or how much you would pay to prevent that much damage. The good news is that, because most deaths will happen a long time from now (probably) the number per year is quite low, around 80.[34] It seems reassuring that killer asteroids are only responsible for a few deaths per year. Given this, we might pay a small premium against this kind of loss. So our asteroid-hunting program would be commensurably small.

That doesn't feel quite right, though. In his book on the dangers humans face, the eminent cosmologist Martin Rees quotes a UK government report: "If a quarter of the world's population were at risk from the impact of an object of one kilometer diameter, then according to current safety standards in use in the UK, the risk of such casualties, even if occurring on average once every 100,000 years, would significantly exceed a tolerable level. If such risks were the responsibility of an operator of an industrial plant or other activity, then that operator would be required to take steps to reduce the risk."[35] Maybe acting like an insurance company is somehow missing the point?

In practice, insurance companies just don't cover huge disasters at anywhere near this scale. Erika Ilves, a consultant who turned into a space entrepreneur when she began to ponder humanity's longer-term prospects, sharply points out in her TEDx Talk that one-off events of complete destruction don't really lend themselves to this approach.[36] After the event human deaths per year go to zero, which sounds great, but that is only because we are extinct. True, we aren't there anymore to care. Nor is there an insurance company to pay out after even a civilization-ending event. My conclusion is that looking at average deaths per year is not the right tool to measure how bad a big impact would be, above some threshold of awful. Extinction events are just in a different category.

So we really do want to know how often this kind of thing happens. But even if we get an answer that sounds safe—once a millennium, say—that's only telling us the odds. A big impact could occur just a few years from now. We just don't know absolutely.

We don't have to gamble. An asteroid impact is the one threat to our existence that we can recognize and avoid for certain. We don't have to be surprised. We can go out and find them before they find us. The wise course of action is to go out and find every asteroid that could threaten us, and then go ahead and move or destroy all those that could ruin our whole day by destroying us. To me it seems worth paying a higher premium and putting some serious work into eliminating that threat.

Guardians of the Earth

When an asteroid is incoming, who you gonna call? There really are people who worry about Asteroid Armageddon for a living: they are the guardians of the Earth. Every two years they get together at a Planetary Defense Conference (PDC) to see what progress has been made. The first one of these I went to was held in Flagstaff, Arizona, near the Barringer Crater, in 2013. We had a field trip to the crater and walked down to the (normally off-limits) crater floor with David Kring, who has spent a lot of time studying this crater. That was quite a privilege.

But the more interesting get-together was two years later, in the spring of 2015, in the beautiful hills of Frascati, above the Eternal City of Rome. It was held at the European Space Research Institute (ESRIN) of the European Space Agency, which sponsored this PDC. I was lucky enough to be among the couple of hundred experts gathered there. Most of the conference was the usual schedule of fascinating technical talks on all the many areas of work involved in saving the planet from an impact by a large asteroid. That is, if anything about saving the planet can be called "usual."

But the attendees also acted out a scenario in which there is an asteroid impact, with updates each day of the conference. In the time frame of the scenario, months or years pass for each day of the conference. All the attendees were assigned roles: scientific experts, world leaders of space-faring countries, world leaders of potentially impacted countries, the media, and the public. As a newbie, yours truly was one of the general public. At first we were a bit awkward doing this—we are techies, after all. But after a couple of days we really got into our roles. Those of us assigned to the general public got seriously angry with the "experts."

We were simulating Armageddon.[37] From my point of view it went like this:

Day 1 (April 2015). An asteroid, 2015 PDC, is discovered. It has a 1 percent chance of hitting the Earth in seven years' time. First estimates are that it is between 140 and 400 meters across. Even if it is at the small end of that range and hits a city, that city would be gone. But while this is a higher probability of impact than any we've seen before, except for one case, Apophis, it's no cause for panic. Nonetheless, astronomers around the world turn their telescopes toward 2015 PDC to learn more.

Day 2 (one year later). Probability of impact rises to a scary 43 percent, which is more or less a coin toss. Impact will be somewhere on a long line from the South China Sea over Vietnam in the east, along the heavily populated Ganges valley of Bangladesh and India, to Pakistan and Iran further west. A hit in many of these locations would cause huge fatalities. The asteroid size is still poorly known, and its composition is completely unknown. Still, it could make a crater 5 kilometers across, 500 meters deep, and would generate a major 6.8 magnitude earthquake. All the space-faring countries are working on deflection plans.

Our panel of experts addresses our "world leaders" to brief them on what could be done. They recommend high-speed impacts to deflect the asteroid.

The public reactions are strong: "What if you miss?" "Are you going to nuke it?" The reply: "Nuclear weapons are explicitly banned in space under the 1967 Outer Space Treaty."

Day 3 (December 2016). Probability of impact is now 100 percent. "What are you going to do?!" demand the heads of state of the threatened countries. In response the United States plans to fire off three large rockets armed with high-speed impactors; Europe, Russia, and China plan to launch one large rocket each. Success requires that at least four of the six find their target.

"What if only a couple succeed?" ask the leaders and the anxious people of the threatened countries. Well, in that case each hit will move the impact point back along the line from the South China Sea toward India and Iran. China is relieved. The people under the rest of the impact path are not pleased. A partial success actually makes things worse for them. News outlets are calling for nukes. Others say, "Why not do nothing?" After all, the tsunami in the South China Sea can be guarded against with concrete seawalls. The impact is still six years away so there is time to build them. The Chinese are not moved by this argument.

The Indian government is particularly unhappy about the result of partial success, as it would direct the impact onto their most heavily populated areas. And India has both a space program and nukes, so when its leaders say that they will launch a "massive observer spacecraft" to see how well the impactors do their job, most observers think that this is a euphemism for sending up a nuke to use as a last resort. Understandable.

Day 4 (August 2019). Three years to impact and the six kinetic impactor spacecraft are ready to launch. They will take seven months to reach the asteroid. If they all miss, the asteroid will hit somewhere in the South China Sea as predicted earlier. If they all hit, the asteroid should miss the Earth entirely. Popular and media opinion is really skeptical that punching the asteroid hard with kinetic impactors is a winning strategy. "Why not nukes?!" we say again, but it's too late now to rethink the ban on having them in space.

In the end, the first impactor breaks the asteroid into two pieces. While the others successfully push the larger chunk away from Earth, the smaller piece continues, delayed only a little. The Indian "observer" spacecraft was knocked out of control by debris from the asteroid breakup, so it was not able to measure the new orbit (or explode a nuke, if it had one).

So now 2015 PDC won't hit in the sea but on land, somewhere in the lower Ganges. On the upside, the impact will release much less energy than the whole of 2015 PDC, just 18 megatons. But that's still 1,000 or so times more than the Hiroshima nuclear bomb. No one in India or Bangladesh is impressed by this achievement.

Day 5 (September 2022). Asteroid hits. It ruins our whole day. In fact it is a major disaster. A large part of Dhaka, Bangladesh, is devastated by the airburst. Dhaka had been the tenth-largest city in the world, with over 15 million inhabitants. Now it is a ruin and millions are dead.

We all come away vowing to do better in real life.

There are lots of lessons to be learned from this exercise, which is of course why we tried it out. An awful lot of new paths for international—in fact, global—cooperation and information sharing will clearly be needed, for one thing. And maybe nukes. Perhaps they don't really need to be banned if we use them not as bombs but as "devices," tools to deflect a killer asteroid?

Every Planetary Defense Conference now includes an exercise like this. And every time, I can now reveal, it is worked out so that we fail and the impact happens. That may sound depressing, but it is set up like that so we can think about every stage of the scenario, including what civil defense we might come up with, such as realistic evacuation plans that may well affect millions of people. In 2019 we obliterated much of Manhattan, but at least we avoided Denver.[38] It turns out that evacuating an island, even with all the tunnels and bridges of New York City, takes time. It was not pretty. Each time the PDC meets, the responses to the threat get better. Each time we learn more of what we'll have to do in a real-life scenario. So even though they are frustrating exercises, they are well worth doing.

Before the Frascati simulation hardly anyone had realized how ill prepared we were. But the exercise was a wake-up call. The response from governments was pretty good. Soon after the Frascati PDC alliances of scientists and government officials started to build up the networks they had already been discussing into effective communications forums. For example, less than a year after the exercise in Frascati, the U.S. government had set up a new Planetary Defense Coordination Office, run out of NASA, to coordinate its efforts. NASA was already working on a plan for the Planetary Defense Coordination Office after a report a year earlier suggested that it should. The office runs programs for detecting and tracking potentially hazardous asteroids and is in charge of working with the Department of Defense, the Federal Emergency Management Agency, other U.S. agencies, and their international counterparts. For now it won't get much in the way of extra funds, but even setting this up so quickly is something of a bureaucratic miracle.

The International Asteroid Warning Network, IAWN, was created in 2013.[39] (It's pronounced "eye-wan," not "yawn.") Its mission is to set up "well-defined communication plans and protocols" so that scientific organizations and governments can coordinate effectively and in time to respond to asteroid threats. IAWN is a clearinghouse for information. For example, the organization keeps a sobering list of close approaches, asteroids that come closer than the distance of the Moon from Earth. They listed 64 in 2018 alone. Similarly, the UN Space Mission Planning Advisory Group (SMPAG) began work at the end of 2013.[40] (SMPAG is pronounced "same page" at the suggestion of Lindley Johnson, who has the great title of "planetary defense officer" in the Planetary Defense Coordination Office at NASA. He wanted to emphasize that that was the purpose of SMPAG—to put all the international bodies on the same page.)

Having some organizations in place is a big step forward. But it wasn't obvious how well they would work together in the event of a real impact. To find out, NASA coordinated a new type of exercise in 2017 involving the relevant parts of the U.S. government, right up to the White House. NASA decided to use the little-known, but real, asteroid 2012 TC4, due to swing by (but not impact) the Earth in October that year.[41] The idea was initiated only in the previous March, but that eight months' warning time is quite realistic. How well would we prepare for impact, using real observatories in real time? The exercise went beyond U.S. organizations. The European Southern Observatories used their Very Large Telescope to find TC4 again in late July. It was confirmed to be the right asteroid in August. Now there were just two months to go. The participants needed to learn more about TC4 to estimate the scale of threat it posed.

Everything did not go smoothly.[42] The bumps led to a lot of lessons learned. One is simply that the details matter. The European Southern Observatory tried to use a thermal infrared camera on one of its Very Large Telescopes to measure the size of 2012 TC4, but missed 2012 TC4 due to small timing errors on the telescope's clocks. That meant that it pointed a little bit away from where 2012 TC4 actually was at that moment. Observations from Hawaii could have helped, but a downed tree had cut off power to the mountaintop at just the wrong time. That ruled out the infrared and meant that the team had to rely on radar to measure the size of the asteroid. But then the weather intervened. Hurricane Maria passed over Puerto Rico at the most unfortunate moment, shutting down the prime radar site at Arecibo Observatory, which is on the island. Scrambling, the team instead managed to get the use of the smaller—though still football field–sized—Green Bank telescope in West Virginia to receive the signal sent from NASA's Goldstone radio dish in California and bounced off 2012 TC4. Finding these glitches in the system is the whole point of the exercise. No doubt real-world trials like this will continue until a well-oiled machine is in place ready to react quickly and efficiently when a true threat arises.

If you find thinking about the fate of humanity a little too much, don't worry. We're about to get to something much more immediate: at last we come to greed.

4

Greed

Our third motivator is greed. Until recently "space" and "greed" didn't often turn up in the same sentence. Space was too noble an enterprise to connect with filthy lucre. In the last few years, though, there has been an explosion of start-ups trying to make profits out of space. Space entrepreneur Peter Diamandis has famously said, "The first trillionaire is going to be made in space."[1] This is a great line and has been often repeated (for example, by Texas senator Ted Cruz and Neil deGrasse Tyson).[2] If someone finds a way to make huge profits out of space, the happy result will be that the costs of space activities will come way down. Then we will be able to afford to visit many more worlds, much more often, and we will be able to find and deflect all the killer asteroids. For me the advent of greed in space is a welcome development.

My collaborator Matt Weinzierl of the Harvard Business School uses the analogy of the Soviet Union to illustrate how space activities over the last several decades have been organized. The Communists were trying to deal with the failures of the market in tsarist Russia. Their government-organized economy went great at first, especially in areas where markets are frequently known to fail: national defense and basic science. Markets fail in these two areas because it's easy to be a free rider on either one of them, as you'll still benefit if others pay. The early space program fit well into that model. No private company was going to develop Moon rockets because there was no business case for them.

In the long run, though, the Soviet system didn't work so well. After a few decades it turns out that the signals that prices provide are really important. Not having them, as they did not in the Soviet Union, means that resources get allocated in wasteful ways. There are no incentives for innovation and cost control in a government-run system, where five-year plans define everything that will happen economically. Without innovation the Soviet Union may have produced record amounts of steel, but more modern steel alloys with carefully targeted markets had no chance to emerge.

Space has suffered from an analogous problem. At the beginning of the twenty-first century NASA needed a new large rocket to replace the Space Shuttle as a way of sending astronauts to the International Space Station (ISS). NASA set up a large program, called Constellation, in the same multibillion-dollar way that it had organized the Apollo program. But at the same time, it set up a small side bet on COTS, the Commercial Orbital Transportation System. NASA administrator Mike Griffin announced the program in 2006. Two companies funded by COTS had cargo flying to the ISS seven years later for a cost to NASA of less than $700 million. Seven years may not sound too quick, but to have two new rockets and two new cargo-carrying spacecraft built from scratch in that time is something that hadn't been done since the early glory days of NASA. Instead the Constellation program fell way behind schedule and was over budget; mercifully, it was canceled in 2010 after just one flight of its *Ares-1-X* rocket. It cost NASA over $7 billion, 10 times as much as COTS. Now, to be fair, Constellation was intended to land people on the Moon and eventually Mars, while COTS was just to get cargo to orbit. Still, Constellation was predicted to cost a total of at least $97 billion had it been completed![3]

This is why greed is needed. As Matt Weinzierl says, capturing the power of the market mechanism is necessary to drive sustained innovation and efficiency in the utilization of space. The goal is not just to make money, but to do so in order to enable all the other things we want to do in space that we otherwise could not afford. *Forbes* magazine used to call itself a "capitalist tool." For space, capitalism is a tool.

The resources of space are vast. How vast? Millions of times what we have accessible on Earth. How can that be since space is mostly, well, just empty space? Yet even the small fraction of space that isn't empty contains resources that dwarf anything we could extract from the Earth. At today's prices they would be worth many trillions of dollars. That's where Peter Diamandis's line comes from. So far, though, both businesses and politicians have paid little attention to these resources, because they seem more like tales of El Dorado than real opportunities.

What are these resources? They have been discussed for years, and are expertly cataloged in John S. Lewis's 1996 book, *Mining the Sky*.[4] They include the helium isotope helium-3 (properly called ^{3}He), which could be mined from the Moon for use in fusion reactors and, from the asteroids, many things, including precious metals, iron to use for construction in space, water for astronauts, and methane for rocket fuel. The iron content of the Main Belt asteroids exceeds the iron reserves on Earth by more than 10 million times. These resources are capable of solving today's fossil fuel–based energy and climate crisis by letting us build solar power stations in orbit that beam gigawatts down to Earth and, separately, shield Earth from the full power of the Sun to relieve global warming.

So any one of these large chunks of rock has got to be pretty darn significant economically, surely? Well . . . maybe. It is true that there is an incredible amount of resources in asteroids. The problem is that, to modify the tagline of the *Alien* horror films, "In space no one can hear you sell."[5] Unless you have customers in space, or you can bring your goods to customers on Earth, those resources are worth exactly zero. And if you do have potential customers, they have to be willing to pay more than it cost you to bring the resources to them. Otherwise you are out of business.

The paeans to the value of space resources have mostly been written by scientists, not businesspeople. And we scientists are pretty financially naïve. Usually we just add up the mass of the resource and multiply by the current price. This approach produces those enticing trillions. For example, the asteroid *Germania* became briefly famous in 2012 when it was estimated to be worth about U.S.$100 trillion. This is huge. The GDP of the entire world for 2012 was only about U.S.$75 trillion.[6] And *Germania* is far from the only apparently super-valuable asteroid. The asterank.com website estimates profitability values for almost all known asteroids. It lists over 100 more asteroids it reckons to have the same value. Asterank doesn't detail, though, how it arrived at its huge valuations.[7]

Those estimates of value, even if they are right, don't close the business case. They are too simple, ignoring the cost of bringing that resource to market, and assuming the existence of a vast market for decades to come with prices that stay as high as they are today. Much more realistic assessments are needed to get asteroid mining going. Nineteenth-century whale hunting has a lesson for asteroid mining. Tom Nicholas and Jonas Peter Akins, both of the Harvard Business School, have pointed out that the only reason anyone would undertake such a dangerous business was the fabulous profit they could make.[8] Does space have a resource as enticing as whale oil? We need something that makes enough profit to get the ball rolling. The profits will have to be as extreme as those of the luckier whalers or gold rush pioneers. And it has to pay off fast.

Will it happen? First off, there have to be profitable resources in space. Then we have to be able to get to enough of these resources to start an industry. That's what we'll look at in this chapter, postponing the questions of how we would do the mining and who we would sell the resources to until later.

Because traveling to space is expensive, for any space resource to be worth bringing back it must have a very high price per kilogram. There are two things out there worth millions of dollars per ton: platinum and water.

The right asteroids do have a lot of one thing that we know we can sell for high prices here on Earth: precious metals. There are six related metals—platinum, palladium, iridium, osmium, rhodium, and ruthenium—that are called the platinum group metals, or PGMs, because they are grouped next to each other on the periodic table of the elements. All six are among the 10 most expensive precious metals. (The others, apart from the obvious gold and silver, are rhenium and indium.)

Platinum is surprisingly useful stuff. The journal *AZoM* reports, "Platinum is integral to the production of about a fifth of all consumer goods. . . . Hard disk drives, anti-cancer drugs, fiber-optic cables, LCD displays, eyeglasses, fertilizers, explosives, paints and pacemakers all rely on platinum. Platinum is also the key catalyst used in fuel cells."[9] In the Earth's crust platinum is about as rare as gold.[10] It is also extremely dense. A ton of platinum would fit into one and a half overhead compartments in an airplane. (Do not try this; not only would it exceed the weight limits, it would be really bad if your cargo "shifted in flight," not to mention the trouble of getting it into the compartment in the first place. Don't check a bag of platinum either; it would be irritating to have millions of dollars rerouted to some distant location.)

Precious metals really are precious. Although prices do fluctuate over the course of a decade, the PGMs sell for millions of dollars per ton. Right now, rhodium is the most expensive, at nearly $90 million a ton. It is rare, though. Palladium (also technologically valuable, and three to five times rarer even than platinum) and iridium are next, at between $30 million and $50 million a ton. Platinum sells for a fairly constant and relatively modest $15 million to $20 million per metric ton and, allowing for inflation, has done so since 1950. It provides the largest payoff per ton of raw material. The two other PGMs, osmium and ruthenium, sell for "only" $10 million to $15 million a ton, and they are less common still. Altogether a value of $36 million per ton of asteroid rock is a decent estimate. That's encouraging.

But why go to space to get platinum? Why not just mine it on Earth as, in fact, we already do? Because some metallic meteorites are five times richer in platinum than the richest ores on Earth, which are in Southern Africa.[11] Why is that? We saw that the metallic asteroids got that way because they come from the broken-up cores of the original small planets of the Solar System, the planetesimals. The planetesimals became molten, due mostly to radioactivity, allowing the PGM elements to dissolve in molten iron (the siderophiles) and flow to their cores. The same process happened on Earth. As a result, most of the Earth's platinum is buried in the iron core, some 3,000 kilometers beneath our feet. Those PGMs are a lot less accessible than platinum from space. Only trace amounts remain in the Earth's crust. That makes PGMs precious.

Didn't we learn, though, that the ores in the Earth's crust, including platinum, came from the asteroids during the Late Veneer era early in Earth's history? So why isn't that ore just as rich as in asteroids? Because, during their impacts, the asteroids kicked up a huge amount of Earth rock that mixed with the asteroid rock, diluting its ore richness. Geological processes reconcentrated the platinum and other siderophiles in just a handful of special places near the surface where we can mine them. So mining the PGM-rich asteroids makes some sense.

The bad news is that even in these asteroids the concentration of precious metals is pretty low. A total concentration of PGMs of about 30 grams per ton is about as good as high-richness meteorites get. Raw metallic asteroid rock, then, contains only about $1,000 worth of PGMs in a ton. We will have to mine a pretty staggering 1 million tons of raw asteroid to get just 33 tons of precious metals, which would be worth $1 billion. That's discouraging. It's certainly going to be a lot of work. On the plus side, we only need to find the right asteroid 45 meters across to have that $1 billion worth of precious metal.

But water is also there and potentially profitable. And water will be a lot easier to extract. Water mining in space is not the most intuitive business venture. Whatever space miners want to sell it has to cost a lot for a small amount, because moving mass around in space is expensive. So it may sound odd that selling water is the leading idea for space miners. Water is not exactly expensive. If Kelly and Zach Weinersmith, authors of the fun book *Soonish*, are right, then the ultimate hipsters will want to drink space water: "I *wish* I didn't have such a delicate palate," says the goateed hipster in one of Zach's cartoons.[12] Still, that's probably a pretty tiny customer base. Space miners won't get rich selling fancy designer bottled water for a few dollars per kilogram. Otherwise, there really is no market for space water on Earth.

In space, though, water is precious. To get a liter of water (weighing 1 kilogram) from the Earth's surface to even a low orbit in space has cost about $20,000 for decades. Getting a liter of water up to a geostationary orbit costs two or three times more, about $50,000 a liter. To put that in perspective, only a few other wildly expensive items really come close: truffles are going at about $10.50 for a gram at the specialty Italian store Eataly, so $10,500 per kilogram. A gram of saffron at the upmarket Wholefoods supermarket goes for $9; that's $9,000 for a kilogram. Both of these are "only" half the cost of water in orbit. The most expensive bottles of wine sell for about the same price as space water.[13] One sold for more: a bottle of Château Lafite from 1787, which sold for $156,000 back in 1985. Allowing for inflation, and for the detail that a bottle of wine is only 3/4 liter, it would come in today at almost half a million dollars a liter! This particular bottle may have belonged to Thomas Jefferson, author of the Declaration of Independence, third president of the United States, and meteorite doubter, so that accounts for the sky-high price. But the whole point of its wild price is its uniqueness. Uniqueness can only be a one-shot deal, and so is not a great model for the space mining industry. One item that outdoes space water handily is jewelry-grade diamonds. A nice engagement ring diamond of one carat (which is just 1/5 of a gram) sells for several thousand dollars. In extremely rare and special cases diamonds can sell for $200,000 per gram![14]

So space water could be sold in Earth orbit for a pretty high price and still undercut water brought from the ground. These extreme prices give asteroid miners hope. They know that it takes a hundred times less energy to bring water from a well-chosen asteroid to a high Earth orbit than to bring it up from the ground, even though it has to travel far across the Solar System. Asteroids have a tiny escape velocity; you could almost jump off them and never come down. Their water is almost already "in space," unlike water coming up from Earth. Does that huge energy advantage translate into a comparable dollar advantage? More basically, will there be any customers in space to buy it? That's a strong "maybe." We do know that in space water is a very useful and versatile resource. Let's look at why.

The big use for space water is to turn it into rocket fuel by breaking down those molecules of H_2O into the hydrogen and oxygen they are made of, using abundant solar energy. Liquefied hydrogen and oxygen are the main fuel for most rockets today. To become liquid, both have to be cooled to very low, or "cryogenic," temperatures of just 20 degrees and 77 degrees above absolute zero (degrees Kelvin) for hydrogen and oxygen respectively. That's quite a complication. Still, the ability to provide rocket fuel up in space instead of dragging it up with you in a rocket seems like a natural win for asteroid mining.

What would the rocket fuel from asteroids be used for? There are several obvious uses for a "gas station" in space. One much-discussed market is the refueling of communication satellites up in their 24-hour orbits. Each of these bus-sized communications satellites is a costly investment of several hundred million dollars, and they have planned lives of 5 to 15 years. If refueling them could keep them operating even a year or two longer it could be worth hundreds of millions of dollars in revenues. Communications satellites need fuel to keep them very precisely in their geostationary orbit. It turns out that, despite the vastness of space, the useful part of geostationary orbit is remarkably small. My good friend and colleague Jonathan McDowell has mapped out their orbits on his website planet4589. org. (The site is named after the asteroid that is named after him.) His data shows that, even though they are 35,786 kilometers above the equator, geostationary satellites are kept within a narrow belt just 2 kilometers thick and 150 kilometers wide, otherwise they won't appear to stay still in the sky above your rooftop dish. None of us want to go out and repoint our dishes constantly, so this accuracy is important to the satellite TV business case. As a result adjustments to their orbits are needed now and then.

Even with refueling, communications satellites will eventually fail. To prevent their signals overlapping communications satellites are spaced out into only about 400 slots; each one is roughly 6,000 kilometers away from the next. The limited number of these slots makes them valuable. Dead satellites need clearing out to make room for replacements. Fuel from space water could be used to kick these deceased communications satellites out of one of these rare geostationary Earth orbit (GEO) slots and up into a "graveyard" orbit, where they will be safely out of the way.

This idea was successfully tested by Northrop Grumman's "mission extension vehicle." Northrop Grumman's *MEV-1* revived a retired communications satellite, *Intelsat-901*, in early 2020. *MEV* doesn't put fuel into an old satellite but instead attaches a new, fully fueled engine to it like a backpack.[15]

Another market will be space debris removal. Space debris is a growing threat to our use of space. De-orbiting space debris, especially from the ever more crowded low Earth orbit zone, could become a sizeable market. The goal is to prevent a chain reaction of destruction from one piece of space junk hitting another at high speed. If a satellite is destroyed in orbit, the result is many pieces of debris that stay in orbit, and they can go on to hit another satellite. This has happened enough that we are already at a point where this debris is undergoing more collisions, creating more space junk, which leads to more collisions, and so on. This vicious spiral is called the Kessler syndrome after Donald J. Kessler who, with his less well-known colleague Burton G. Cour-Palais, pointed out the problem in 1978.[16] The situation was made much worse by anti-satellite weapons tests by China in 2007 and India in 2019. Each of them left hundreds of bits of space shrapnel to deal with. This runaway growth of space debris could make low Earth orbit unusable, at great economic cost. Jonathan McDowell catalogs all objects launched into space.[17] He estimates that it would take several hundred tons of water propellant to remove all the space junk. A European consortium launched a test satellite, aptly named *Remove-DEBRIS*, from the ISS in 2018 to test methods of grabbing space debris: with a net, with a harpoon, and with a "drag sail." All three methods worked.[18]

Another use for a fuel depot in orbit is refueling upper stages of Moon- or Mars-bound rockets. As Mars rockets leave only every 26 months, when Mars comes to a convenient place in its orbit, they would form a rather boom/bust cycle for space-derived fuel. Moon rockets are rare right now, but look set to be a growing market if the dozen or more lunar landings planned for this decade start a new phase of intense lunar exploration.

An easier type of asteroid rocket fuel to handle than hydrogen and oxygen may be methane. This could be made from the large amounts of organic compounds on carbonaceous asteroids. Methane is a rocket fuel, but has not been used much yet. It has the advantage that it doesn't need to be stored at such low temperatures as oxygen and, especially, hydrogen. Several new rockets will be powered by methane-burning engines: the SpaceX *Super-Heavy/Starship,* the ULA *Vulcan,* and the Blue Origin *New Glenn.* The last two will use the same BE-4 engine. *Starship* is explicitly designed to be refueled in orbit.[19] SpaceX is planning on using methane precisely because it can be manufactured in space, in particular from the thin Martian atmosphere, which is almost entirely carbon dioxide. Asteroid methane could later be a feedstock for making more complex organic molecules in space.

Then there's life support, something astronauts are quite keen on. Space tourists will like it too. Water is needed for drinking but, even more fundamentally, for breathing, once it is broken down to free up the oxygen. If a space station is to be self-supporting, water will also be needed to grow food. Since a modest green salad portion (100 grams) would presently cost $2,000 to bring up from the ground, growing your own in space could make good economic sense too. Using water to maintain life is not much use unless there are people in orbit. Is there a demand? Not right now. Keeping three astronauts on the ISS takes only three tons of water a year.

Could there be a demand? Yes. In fact, this could become a fast-growing industry. We'll get to some of the possibilities later on.

A less obvious use for space water derives from the fact that the hydrogen in the water is great at shielding the inside of space stations from the deadly galactic cosmic radiation. On long stays in space humans absorb an unhealthy carcinogenic dose of radiation, and a layer of water around their living quarters would cut this down to a safer level. As they will need water anyway for life support, why not wrap it around them for protection? The total they'll need to be safe is in the thousands of tons range, as the layer would have to be several meters thick. There's no way that is affordable at $20,000 per kilogram; it would cost over $20 billion! Asteroid water might solve the problem. Cutting the cost ten- or twenty-fold would be needed to make the shielding something close to affordable. That is not something asteroid miners can do yet. But as a way to expand the market later on it looks good.

Water is a lot easier to extract from rock than platinum. Asteroid water will not come as blocks of ice though. There is a slight chance that some might, but most near-Earth asteroids have spent too much time close to the Sun. Instead, the water in asteroids, based on what we find in meteorites, is chemically bound into clays and similar rocks as hydroxyl (OH). It can take a lot of energy to extract this water. But there's a lot of free sunlight in space that could be concentrated to heat up these rocks and make them give up their water. Water extracted that way won't satisfy U.S. Environmental Protection Agency standards for drinking water, and so it will need purifying if we want to make tea with it. Worse, Alessondra Springmann, while a PhD student at the University of Arizona, showed that warming up meteorites allows some quite nasty elements in the rock to move around and contaminate the water. Mercury, for example, is not only injurious to your health, it could cause metal failure in your rocket fuel manufacturing equipment.[20] At least water used for radiation shielding doesn't have to be clean and pure.

What about more exotic resources? The idea that we might find truly novel materials in space is common in science fiction. Could there really be strange new materials to be discovered in space? The answer is a very definite "Yes!" Hunting down asteroids could well result in remarkable new materials that could be technologically valuable. Some of these may eventually form the basis for new industries. Could they be enormously valuable? In the movie *Avatar*, people travel for decades to another star to bring back prized supplies of unobtanium, a room-temperature superconductor found only in space. Unobtainium is an old joke name used in physics laboratories for something you can't have. When the professor asks, "Why can't we do X?" the graduate student replies, "We could, if only we could get some unobtainium!" Hilarity ensues. In our universe, the asteroids are a more reachable place to find strange new materials than other star systems.

What we won't find are new elements. An element is made of atoms all of the same kind with the same number of protons. By the late twentieth century every one of the naturally occurring elements had been found. We can be sure of this because we have found atoms with every number of protons, from 1 to 94. There are no gaps to be filled. Beyond these are 24 elements with even more protons. They were created by people using nuclear reactions. (That's why the reactions are called "nuclear"—they involve changing the number of protons and neutrons in the nucleus of an atom.) Most of these new elements last only a tiny fraction of a second before spontaneously decaying into the naturally occurring elements. We know all the elements.

But this doesn't mean there's nothing new to discover. The number of combinations of the elements into compounds—molecules, crystals, and variations on these—is effectively infinite. Just two types of atoms can come together in combinations we cannot predict, even with the best imaginable computers. The number of possibilities to explore is just too vast. Fortunately, we don't have to. The asteroids can tell us. There really is a lot of potential for finding new materials in the asteroids that are not found on Earth. Some may have useful and valuable properties, even if we can't predict just what they will be. That is an expectation based on the fact that new materials have historically found new uses. It may be wrong, but I doubt it.

Jöns Jacob Berzelius found the first meteorite mineral, schreibersite, in 1832. Meteorite minerals are those that were found first in meteorites. Only a small minority of them have yet been synthesized or found later on Earth. Berzelius may not be a household name today, but he was one of the four great founders of modern chemistry. He invented the "H_2O" shorthand for molecules. His notation makes it much easier to see what is happening in reactions because H_2O is much easier to take in than "di-hydrogen monoxide." More fundamentally, his notation embodies his finding that elements combine in whole numbers. (That way you always get H_2O, not $H_{2.5}O$, for example.) That gave strong support to the (then new) idea that all chemical compounds are made of atoms of a small number of elements. Identifying which chemicals were elements was naturally important to make this idea stick, and he discovered four of them, including silicon. He also had a side interest in mineralogy that led him to find schreibersite. That's some side interest! He was later honored by having the mineral berzelianite named after him. He wasn't always on the right side of arguments, though. He denied that chlorine was an element, and proposed that life depended on a "vital force," neither of which was a winning idea.[21] That doesn't mean he wasn't a great scientist. Science is a messy business when difficult problems are being worked out. None of us are on the right side of everything.

My talented student Nina Hooper combed through dozens of publications and tabulated over 70 meteorite minerals. Very few of them have been investigated carefully for interesting properties. That's because most of the samples we have are tiny, a fraction of a millimeter in size, similar to the thickness of a human hair. That makes them hard to handle and experiment on. Even so, a few already show signs of peculiar properties that we may someday exploit. Their small sizes will also make them hard to mine and to exploit, of course. The valuable ones may be only those that occasionally can be found in larger chunks, if they exist.

Because only the toughest pieces of asteroids make it to the ground as meteorites, we may be missing many more meteorite minerals. If they form in the weak matrix of an asteroid and burn up as the rock fragments during its searing trip through our atmosphere, we will never know about them. If we could go out to an asteroid and search it for meteorite minerals before they suffer that trip, we may find many more, with a correspondingly better chance of discovering something extremely valuable.

Why would there be novel minerals found only in space? Because the Earth has tectonic plates that are constantly colliding, creating mountain ranges and sinking beneath them. As a result, old rocks are hard to find here. Every couple of hundred million years the crust of the Earth, on which we live, is mostly recycled. That means that there are very few rocks here that are nearly as old as the Earth, and they are difficult to locate. The Nuvvuagittuq Greenstone Belt, which is on the coast of the Hudson Bay in Northern Quebec, is one of the oldest exposed terrestrial rock formations known, at about 4 billion years old.[22] It is just half a kilometer wide by a few kilometers long. Asteroids, however, have no plate tectonics. In fact, unlike the planets, nothing much has happened to them since the epoch of the Late Heavy Bombardment, so rocks dating back to the formation of the Solar System abound on them. Okay, but why would ancient rocks be different from those made much later? The key is that in space there are conditions that never applied on Earth. Some of them we can never reproduce in laboratories, even if we put those labs in space. Those extreme conditions can lead to new materials.

Asteroids are natural experiments in material science. There are at least two ways they are special: time and violence. We can duplicate the vacuum and cold of space in the laboratory on Earth and we can also duplicate the weightlessness of matter in asteroids in orbit. But deep time and extremely violent collisions are beyond us.

We cannot duplicate the 10 million years over which the pre-solar nebula condensed into planets and asteroids. If there are exotic molecules or crystals that grow that slowly we could not make them. It's really hard to get a grant that lasts that long. Graduate students tend to want to get their PhDs a little quicker too. The best-known example of a slow-growing crystal is the Widmanstätten pattern (figure 5). This is an alternating leaf-like pattern of two iron-nickel alloys, taenite and kamacite. It is often found in the metallic meteorites. This pattern grows by millimeters in a million years in the metal core of a planetesimal as it slowly cools.[23] Try doing that in the lab! The Widmanstätten pattern itself has no practical use though.

Tetrataenite is one meteorite mineral that has been investigated quite thoroughly. This is another nickel-iron alloy, with traces of copper and cobalt. Those two extra elements, and the structure they form when cooling very slowly in the core of a planetesimal, make a big difference. Tetrataenite has unusual magnetic properties. It is unusually hard to change the direction of its magnetic field. (Technically, it has high coercivity.) That

Figure 5: An example of the slow-growing Widmanstätten pattern in the Toluca nickel-iron meteorite.

might be interesting for computer applications. Members of a research group in Japan are seeing if they can make tetrataenite by some new process. If they succeed, then it is still important that it was found in meteorites because otherwise no one could have guessed that it was possible. If they can't synthesize it, then maybe one day there will be an industry mining tetrataenite from asteroids. This mineral forms best in the asteroids that cooled the slowest, so identifying which those are would become important. There are a few other meteorite minerals that are promising: josephite is magnetically similar to tetrataenite, and panguite has novel conductive properties.[24]

We already have meteorites, of course. Why not just mine them? Because there are only about 500 tons of metallic meteorites in museums today.[25] That's not enough to support an industry if one of them were to become a technological must-have. Also, curators won't be keen to have their entire collection destroyed to find a few specks of meteorite mineral, however valuable it may become. We'll have to go to space to get the valuable stuff.

All the above discussion of exotic materials has involved only inorganic molecules. The carbonaceous meteorites show us that asteroids contain a large number and variety of organic molecules, as we saw when we discussed the origin of life. (Jöns Jacob Berzelius was also the first person to introduce the idea of organic and inorganic molecules.) Among these organic molecules are many amino acids, the most basic building blocks of life on Earth. Some of them were unknown until they were found in a meteorite.

Our samples of organic material from asteroids are extremely limited. There's far less carbonaceous meteorite in the world's collections than metallic meteorite, just 173 kilograms. Worse, only a small number of carbonaceous meteorites have been seen to fall. And only a handful of them were found as fresh falls in locations that may preserve them well. The best and most famous example is the Tagish Lake meteorite.[26] This fell in Canada in 2000 onto a frozen lake in the Yukon and was partially collected with almost no contamination or alteration once it had reached the ice on the lake.

We have this sample only thanks to a series of lucky breaks. First, the lake was frozen, so the meteorites didn't just plop out of sight; then the black rock of the meteorites stood out against the snow; then a local man found the meteorites just after they fell, and thought some scientist might want them, and so picked them up and put them in a cooler on his front stoop. The very next day it snowed again, covering up any other meteorites on the lake. A few were found later, but they had sunk into the ice and are less pristine. Only about 10 kilograms were collected in all. Estimates are that this is only 1 percent or so of the incoming rock. If the man had decided to take them indoors they would have warmed up overnight and would have lost many of the volatile materials they had carried for billions of years. Once they were warmed up at NASA's Johnson Space Flight Center in Houston, that's what happened, but the gases were captured and analyzed. To find many novel meteorite minerals with such a limited supply we'll have to be very lucky. Or we can go to some carbonaceous asteroids out in their native orbits and collect samples there. The beginnings of that exploration are under way.

Despite the amino acids, sugars, and possibly proteins already found, the full range of complex molecules in asteroids may well be more complex than we can find in meteorites.[27] More delicate and more complex compounds may be found by directly sampling asteroids in space. Given the complexity of organic chemistry, unexpected compounds may be found in them. Some of these may have technological or medical uses. At least one example of potentially medically useful meteoritic material is already known. Isovaline is a rare amino acid first found in the carbonaceous Murchison meteorite. Isovaline has been reported to have promise as a painkiller as, unlike opiates, it cannot cross the blood-brain barrier.[28] There's a lot of work to do to prove that promise though.

The other property available in space we cannot duplicate in the lab is the enormous speed of collisions between asteroids, about 88,000 kilometers per hour. At that speed you could circle the Earth in half an hour. That kind of violent collision is destructive, breaking up the asteroids into rubble piles, but it can also be creative, building new types of material that we can't duplicate on Earth precisely because the speeds are so high. We have clues that these violent collisions can make some strange compounds.

A great example of what violent impacts can create is natural quasicrystals. Quasicrystals are a fairly new discovery. The 2011 Nobel Prize in chemistry was awarded to Dan Shechtman "for the discovery of quasicrystals," which had been made in the lab in the 1980s. Until quasicrystals were found, every solid was either amorphous (like metals), or a crystal (like quartz). All crystals had a repeating pattern that could be shifted sideways or up and down and find a perfect match, like wallpaper. In the 1960s mathematicians found that other types of patterns existed that had a well-defined structure but did not allow this translational symmetry, as it is called. However, no such patterns had ever been found in any minerals. Shechtman, unaware of this (rather arcane) mathematical result, was doing routine X-ray analysis of materials when he found a tenfold pattern in the data that should not have been there for any normal crystal. After more work, and some natural hesitation, he announced this discovery—the first quasicrystal. A rush of others soon followed. We can now make many types of quasicrystals on Earth, but there are so many possible forms they could take that it's hard to cover them all. It would be handy to let nature do the work for us. That's not so easy to do, as we don't find quasicrystals occurring naturally in any Earth rocks.

Instead, the first quasicrystal to be found "in the wild," rather than being created in a lab, came from a meteorite found near the Khatyrka River just north of the Kamchatka Peninsula in easternmost Russia by the Bering Sea. This was no chance discovery, but the end of a careful decade-long hunt by Paul Steinhardt of Princeton University that is part detective story, part adventure.[29] The case that the quasicrystal had been found in a meteorite was controversial at first. For one thing, it was quite perfect, equal to the best that could be made in the lab. Surely that could not be natural? It also included metallic aluminum, which is really unusual on the Earth's surface because it oxidizes very quickly in air. It would have needed super-high pressure and temperature to form. It likely formed in just the sort of high-speed collision between two asteroids that would reduce them to rubble. Those conditions can also be found deep in the Earth's mantle, so whether it was a space rock was still ambiguous. Isotope ratios of the oxygen in the rock solved the dilemma. They lined up perfectly with those of carbonaceous meteorites and not at all with those of mantle rocks.

The International Meteoritical Association has now accepted that it really is a meteorite, and it is named it Khatyrka, after the river.[30] The quasicrystal is made of aluminum, copper, and iron and is called icosahedrite. Icosahedrite was already known as lab-grown quasicrystals. Nonetheless, it is remarkable that it forms naturally only in a space rock. Before this discovery it was even unclear whether any quasicrystals could be stable. The Khatyrka quasicrystal survived in space from the earliest times of the Solar System, so that question was settled. After seven more years of painstaking work the Khatyrka meteorite finally yielded two more quasicrystals; this time one of them was new to science. Quasicrystals don't yet have any technological applications. There are many teams pursuing a wide range of ideas on how their special properties might be useful.

The violence of asteroid collisions may not stay unique for long. The Sandia National Laboratories' Z machine, in New Mexico, can now accelerate small samples to this kind of speed. For now, this is the only facility on Earth that can do this.[31] The Z machine is mainly used for nuclear fusion research, not for exotic mineralogy. It is a sign, though, that we may not have to rely on asteroid collisions for this kind of science much longer. Deep time, however, will always remain unique to asteroids.

You might think that with almost 200 years of research since Berzelius's first discovery, we must have found all the meteorite minerals by now. Not at all. By 1950 only 5 had been discovered. Of the 70 or so meteorite minerals we know of now, about half were found within just the past 20 years.

And two-thirds of those were found by one scientist, Chi Ma, and his team at Caltech, only since 2009.[32] This suggests that there are plenty more to be found. As our tools and samples improve, there are likely to be dozens, maybe hundreds of new meteorite minerals to be found.

It would be remarkable if no meteorite minerals had peculiar properties that make them valuable for some technology. Once we know about them, some may be synthesizable in Earth-based laboratories, others in space-based microgravity laboratories. A few may point the way to whole new classes of materials for condensed matter physicists. And some may resist forming under any conditions we can create. If these last ones are useful, they will have to be mined in space. If they are valuable enough, they may be what eventually drives the space mining industry, rather than the bulk materials—water, methane, and PGMs—that are the focus of present efforts.

Still, for now water and precious metals look like the way to get started with space mining. We'll need much more revenue, though, to have a space industry that is an important part of the world economy. What will come after these initial pump primers? Meteorite minerals are something of a long shot. Can we see something else that could grow the in-space economy to truly large scales? Yes—tailings. Tailings are all the detritus left over after you've extracted the valuable ore from a mine. They are also called gangue. On Earth they have virtually no value. But there are no tailings in space; it costs too much to get anything up there. Anything we can control in orbit is going to find a use eventually, and so will become ore.

The tailings from mining precious metals will be almost pure nickel steel. Precious metals are so rare that if we mine even a PGM-rich asteroid that is football stadium–sized (100-meter diameter) for precious metals, we'll only extract a total of about 50 tons of them, leaving behind about a million tons of virtually pure nickel-iron. If we could move it somewhere that we could use it, then it would be a great resource. Not incidentally, doing so could make us a lot more profit for a marginal extra cost.

The problem is, how do we move huge masses of steel around in space? Large masses are hard to move in space. That's why we have so far concentrated on products that are worth a lot per kilogram. Moving a million tons sounds truly unlikely. The answer comes from reexamining our assumptions. To get a good return on investment we need to provide a profit in just a few years. That forces us to use trajectories that are quick to traverse, but that use more energy than is strictly needed. There are other orbits that require very little energy to move an asteroid to the Earth-Moon system. These are trajectories on "invariant manifolds" that take advantage of multiple planets to minimize the energy needed. Their drawback is that they are slow, taking years rather than months to move to Earth orbit. But if we have already made a handsome profit from extracting the platinum group metals, and if the marginal cost is low enough, then we won't mind waiting, because anything extra that we can sell is just icing on the cake.

A million tons of construction steel is a lot. It's enough to build about 20 copies of Dubai's Burj Khalifa skyscraper, the tallest building in the world. The ISS, the largest spacecraft ever built, comes in at only 419 tons. What could we do with a million tons of steel in space? There are a few large enough ideas: building hotels, solar power stations, and geoengineering.

What about the "dumb rock" tailings from water mining? Is there any use for that? Each ton of water will create only 10 tons or so of rubble. Still, that adds up. Extracting $1 billion worth of water will create at least 200 tons of tailings. Silicate rock is also pretty good as shielding to protect people from powerful cosmic rays, if you have enough of it. One place those tailings could be used is a Mars cycler. Mars cyclers are bodies (which could be spacecraft or asteroids) in carefully chosen, highly elliptical orbits that continually cross the Earth's orbit and then Mars's orbit at just the right times so that Earth and Mars are right there every time as they cross. These special orbits were discovered by Buzz Aldrin of Apollo 11 fame, and two colleagues, Dennis Byrnes and James Longuski.[33] Travelers would take a small taxi spaceship to rendezvous with the cycler near Earth, ride the cycler to Mars's orbit, then descend to Mars's surface in a local Mars taxi spaceship. Once a cycler is set going in its orbit there's no need for any fuel. That means a cycler's living space can be wrapped in the thousands of tons of water or rock needed to shield the travelers within from carcinogenic galactic cosmic rays. Since we can mold the asteroid into concrete, the travelers could also have a lot more living space for their five-month trip. Maybe they could even enjoy partial gravity by spinning the whole cycler. Asteroid rock is the only cheap enough source for that much shielding. This is not a very distant prospect. Damon Landau and Nathan Strange of the Jet Propulsion Laboratory (JPL), with James Longuski and Paul

Chodas, have compiled lists of near-Earth asteroids that we already know about that could be easily nudged into Mars cycler orbits using today's rocket technology.[34] Of course, this has to be done carefully as there's a fine line between zipping by close to Earth on your way to Mars, and actually hitting the Earth head-on!

There are huge resources in space. They have a host of potential uses. Will there be profits from space, though? In other words, can mining make for a profitable business case, satisfying our greed? That depends a lot on bringing down the cost and the risk of working in space, and on the creation of a demand for materials in space.

Next we look at means—how we might explore the asteroids for all of our motives: love, fear, and greed.

PART

III

Means

5

Love: Doing Science with Asteroids

You are fully motivated now. Love, fear, and greed drive us to the asteroids. Like a detective, though, we need to discover more than motive to solve the crime; we also need to determine means and opportunity. Let's start with the means.

First, love. With so many Big Questions tied to the asteroids, they must be a big field of science, surely? Not really.

Every three years the world's biggest meeting of scientists from all over the world who work on asteroids, comets, and meteorites is held. Naturally the sites where the conference takes place are all over the world too. The 2014 Asteroids, Comets, and Meteorites meeting was held in Helsinki. It was the first one I had been to, and so I was amazed and honored to have an asteroid named for me announced at the end of the banquet, 9283 Martinelvis. I had no idea this was coming. If I had, maybe I'd have had less wine! Of course, many others got their named asteroids that night, mostly the newly minted PhDs in the field. Getting your own asteroid is like graduating. If it were the Mafia it would be like becoming a "made man" (but without the homicide; they take care not to name any potentially hazardous asteroids after living people!). It was great to become an acknowledged, albeit junior, member of the field. The other surprise for me was to see that this worldwide gathering had only about 400 participants. To put that in perspective, the adjacent fields of astronomy on the one side and geology on the other respectively pull in 10 and 50 times as many attendees for their big meetings.[1] So asteroid science is a pretty small pursuit, at least for now.

How is asteroid science done? There are three very different ways to go about investigating the asteroids. First, we can use the techniques of astronomy to tell us about asteroids, using telescopes on mountaintops on Earth. Second, we can send out a space mission to an asteroid to poke at it up close and personal, and perhaps bring a sample back to analyze carefully in laboratories on Earth. Third, we can let asteroids bring us their own samples as meteorites.

Why not just go there with spacecraft probes? Isn't that obviously the best approach? Yes, it is; but each such probe costs hundreds of millions of dollars, even a billion. It makes sense, then, to do a lot of preliminary work using cheaper means—telescopes and meteorites—despite their limitations.

I'm an astronomer, so I'll tackle what astronomy can teach us first. If you're looking through an Earth-based telescope, an asteroid is just a moving dot of light in the sky. That's why Herschel named them asteroids— "not a star." We know what stars are made of, thanks to spectroscopy, and how far away they are, courtesy of astrometry—the careful measuring of their positions and whether they have another star or planet orbiting them, thanks to photometry, accurately monitoring their brightness. These three techniques can tell us about the asteroids too. Spectroscopy tells us the minerals on their surface. Astrometry tells us their orbits and, in the most favorable cases, their masses. Photometry tells us how large they are, how fast they spin, and something about their shape. Occasionally we see that an asteroid is about to pass in front of a bright star and block out its light for a while. (This is called an occultation.) As we know from its orbit how fast the asteroid is moving, the length of this odd form of an eclipse also tells us how big the asteroid is.[2]

But there are a few wrinkles with using these tools for asteroids. The first is that, while stars do change their positions on the sky, they do so very slowly. Asteroids, on the contrary, move a lot. That means that to discover them you can't just point your camera at some place in the sky once as you can for stars or galaxies. Instead, you have to take at least two snaps in quick succession to pick out your moving asteroid. In practice you need to keep on going back to look again several times to trace your asteroid's path in the sky. That way, you can see which way and how fast it's going and work out its orbit. Often some other asteroids will have wandered into that patch of sky since you last looked. You need to track your asteroid well enough that the next time it comes around near to the Earth (its next apparition) you'll be able to find it again, even if that is several years later, so that you will know that it's the same asteroid you found last time.

Discovering asteroids is now done mainly by a couple of telescopes designed to image large areas of the sky at once: the Catalina Sky Survey based near Tucson, Arizona, and Pan-STARRS-1 on Haleakala, the tallest mountain on the Hawaiian island of Maui.[3] Between them they find about 2,000 near-Earth asteroids a year and many thousands of Main Belt asteroids. These two telescopes are so sensitive that the near-Earth asteroids they find are already quite faint. They naturally tend to find them when they happen to be closest to Earth and so appear at their brightest in the night sky. Then they rapidly get fainter as they move away from Earth.[4] That has meant that a sizeable fraction of their discoveries fade away too fast to measure the arc of their orbits well enough to find them next time around.[5] They are effectively lost.

The second wrinkle is that rocks have much more boring spectra than the hot gases that stars are made of. An isolated atom in a gas produces a relatively simple pattern of bright or dark lines in its spectrum. That's how Bunsen and Kirchhoff were able to identify what was burning in Mannheim, and then to find out what the Sun is made of. But atoms trapped in solid materials produce extremely complicated patterns of lines that blend together, mushing out their signatures into broad bands. These bands tell us roughly what types of minerals there are in the rocks, but not the details. Nevertheless, this is enough information to get a first idea of whether an asteroid is mainly "just" rock, or whether it is carbonaceous and contains water in the rocks, or whether it could be solid metal. Even these features, though, are quite subtle; we need really well-exposed spectra to discern them.

The third wrinkle has to do with how bright an asteroid appears to be. Ideally, we could tell how big an asteroid is from how bright it appears at some particular distance from us. This works for stars, if we know their temperature. But asteroids only reflect sunlight, so to know how big they are we first need to know the fraction of sunlight each of them reflects, their albedo. That's a tricky property to measure.

We also need to know the asteroid's phase. Like the Moon, asteroids have phases: sometimes we see them "full." That's when they are directly behind the Earth and in line with the Sun (at opposition). (They are usually too far away to be in the Earth's shadow.) Sometimes we see them half full (at quadrature). You'd expect that the half Moon or asteroid would be half as bright as the full Moon or asteroid. In fact, the "half" phase is 10 times fainter than the "full"! The main reason for this is that near the day/night line the Sun is shining down almost at the horizon. It is dawn or dusk there, after all. That means that a small amount of sunlight is spread over a large area of surface. (Think of the long shadow you cast at sunset.) Instead, when we view the "full" phase we are looking at midday over most of the area we see. We have to take our viewing phase into account to work out just how much sunlight the asteroid is reflecting, but this is not too hard.

We can get a first measure of their masses from the minority of asteroids that are actually two asteroids orbiting one another—the binaries. This has yielded a surprise: most asteroids are no denser than water, while the rock they are made from is five times denser.[6] They must be full of holes! That might be heaven for zero gravity spelunkers but will create some tricky problems for doing anything with asteroids. These gaps are called voids. Perhaps this hole-i-ness of asteroids shouldn't be surprising. The violent early history of asteroids kept breaking them up, so that many of them are just rubble piles held together by their own very weak gravity. Odd-shaped pieces of rubble are not going to fit back together like bricks in a wall. There's bound to be a lot of gaps. Are these 15 percent of asteroids typical of the rest? We presume so, but we can't be sure.

One neat thing we can do quite easily is to measure how elongated an asteroid is from seeing how much it changes in brightness as it spins. Some asteroids can be extremely long and thin. Asteroid (216) Kleopatra looks like a dog bone (see figure 1). When it is side-on to us it looks much brighter than when it is end-on. With really high-quality records of how an asteroid changes in brightness with time (its light curve), it is possible to get quite a lot of detail about its shape using the techniques of tomography. Tomography is the medical imaging trick of taking X-rays of your body from lots of directions at once. We look at an asteroid only from one direction but, conveniently, the asteroid rotates for us. In both cases you can work out the 3D shape of what you are looking at.[7] Occultations happen when an asteroid passes in front of a star; how long the star winks out tells us how big the asteroid is, because we know how fast the asteroid is moving. Very few sizes have been measured this way as yet.

Kleopatra is a somewhat special case. It is one of the thousand or so asteroids that have come close enough to the Earth that they can be probed actively by bouncing a radar signal off them.[8] When an asteroid does come near enough to use radar it returns spectacular data. It is no longer just "not a star" but shows its size and shape, its rotation rate, whether its surface is smooth or rough, its albedo to radar—which is especially high for metallic asteroids—and whether it has moons. Radar signals also measure an asteroid's distance and speed to extraordinary precision, refining its orbit to clearly rule out (so far) an imminent impact with our planet. Astronomers can only be envious. We passively collect radiation sent out by distant stars and galaxies. The most distant celestial bodies probed by radar from Earth are the larger Main Belt asteroids. The returned pulse becomes too faint, too fast to go further out. That's because the strength of the pulse that reaches the asteroid gets fainter as the square of its distance from us, and the same happens to the radio pulse reflected by the asteroid on its way back to us. So the same asteroid placed twice as far away will return a signal that is 16 times weaker; 3 times further away it will be 81 times weaker. Most near-Earth asteroids, being small, intercept only a small fraction of the pulse that we send out, which further weakens the signal they reflect back to us. To be detected, even with the giant, but now collapsed, 300-meter diameter Arecibo radio antenna in Puerto Rico, they typically have to come within about 20 times the distance of the Moon from Earth.

We'd like to know more; but achieving that remotely from mountaintop telescopes doesn't look possible. Accurate sizes and masses of most near-Earth asteroids are beyond reach. Having selected the one in a thousand asteroids that looks promising, we'll have to visit each of them to assay their value more accurately. That necessarily means space missions.

Asteroids were not the first choice of Solar System explorers in the Space Age. But gradually they have gotten the attention of more scientists. At last, in the twenty-first century, they have started to receive some serious respect. There are two missions to asteroids under way and two more are being built. But even after they have been given the green light, missions take years to build and years more to reach their targets. As of 2019 eight spacecraft missions had visited within 1,000 kilometers of an asteroid; sometimes a mission manages to visit two.[9] That's about one asteroid per year, which is a big step up from zero before 1999.

NASA has a mantra for the four steps in the sequence of missions to a planet: fly by, orbit, land, and sample return. Each step in the sequence produces far more information than the last one, but is also more difficult and so costs more. The same sequence is likely to be followed by space miners.

The first asteroid flybys were made by missions on their way to somewhere else. NASA nudged the trajectory of its flagship *Galileo* spacecraft while it was on its way to its main objective, Jupiter.[10] The idea was to take advantage of *Galileo*'s passage through the asteroid Main Belt. First *Galileo* flew by Gaspra in 1991 at a distance of 1,600 kilometers. Gaspra is a stony-type asteroid and is fairly large, about 10 kilometers across. Two years later *Galileo* flew past the larger, roughly 30 kilometers across, Ida but at a little greater distance (2,390 kilometers). Ida had a surprise waiting: a small moon, just 1.5 kilometers across, orbited it. This moon was later named Dactyl, after the mythological Dactyls who lived on Mount Ida in Crete. This was the first asteroid moon to be found, though now we know of quite a few others. Not to be left out, the European Space Agency sent the famed *Rosetta* comet mission past another Main Belt asteroid, Lutetia, in 2010.[11] Lutetia is huge compared with Ida or Gaspra, about 100 kilometers in diameter, and it's relatively round.

The first spacecraft to get into orbit around an asteroid was NASA's *NEAR Shoemaker* mission.[12] The spacecraft was named after Eugene Shoemaker, who showed that the Barringer crater was created by a meteorite impact. (The "NEAR" part stands for "Near Earth Asteroid Rendezvous," as opposed to the Main Belt asteroids visited briefly by *Galileo*.) *NEAR Shoemaker* arrived at Eros in 1999 after a three-year voyage that flew past the asteroid Mathilde on the way. Going into orbit means carrying more propellant because you have to fire your rockets to slow down. In a classic demonstration of how much harder it is to orbit something, *NEAR Shoemaker* nearly missed getting into orbit around Eros when a mystery malfunction sent it spinning. Only after 24 hours of hectic work and with zero reserves of fuel did it make orbit. Eros was the very first near-Earth asteroid to be discovered, back in 1898. Until then all the asteroids discovered had been in the Main Belt. It is a 20-kilometer-sized stony asteroid. Eros had a historic role in measuring the scale of the Solar System. Just three years after its discovery it came close enough to the Earth that two telescopes on opposite sides of the Earth saw it in different positions against the background stars, just as your finger appears to move against the background if you look at it first with one eye and then the other. That, combined with some other data, let astronomers work out how far away it was.[13] From that they could tell how far away the Sun was, setting the scale for the whole Solar System.

Being in orbit meant that *NEAR Shoemaker* could return many more images than *Galileo* had, and gave it time to use specialized instruments that needed long exposures. Eros turned out to have a lumpy potato shape, and it was all covered in loose-looking rubble of small rocks and pebbles. Just seeing that detail was a big step forward. The papers describing the results carefully talked about the *geo*logy of Eros. At the time it was conventional for geology to refer only to Earth rocks. Selenology was used for the Moon, while areology applied to Mars. (Ares is the Greek name for Mars.) Larry Nittler, one of the papers' authors, explained to me that extending this method of nomenclature to Eros with "erotochemistry" was too much for the polite editors of the journal *Icarus*. So now *geology* is used for any planet, moon, or asteroid. It's simpler, and less awkward, that way.

At the end of *NEAR Shoemaker*'s prime mission in 2001, it had all but run out of propellant. The NASA controllers said, effectively, "Hey, why not?" and had it touch down on the surface, so they officially scored the first landing on an asteroid.

NASA followed up on this success with the more ambitious *Dawn* mission.[14] This time the targets to orbit were two of the largest asteroids: Vesta (525 kilometers diameter), which it reached in 2011, and Ceres (950 kilometers diameter), where it arrived in 2015. Ceres is the largest asteroid, the first one discovered 214 years earlier by Piazzi. Together these two big Main Belt asteroids contain the majority of the mass of all the asteroids.

Stepping up in difficulty, the first sample return from an asteroid was by the Japanese space agency JAXA (Japan Aerospace Exploration Agency).[15] Its *Hayabusa* spacecraft first orbited and then landed briefly on the near-Earth asteroid Itokawa in 2005. Itokawa was named after the early Japanese rocketeer Hideo Itokawa. Itokawa is relatively tiny: just about 300 meters long. But it is much easier to get to than 98 percent of all other known near-Earth asteroids. It is easier to get back from too. *Hayabusa* had a number of mishaps on its journey, but still managed to land briefly, collect a sample, and return a milligram of surface dust to Earth. Yet that milligram was the first sample ever returned from an asteroid in pristine condition. Unlike almost all meteorites, we also knew where it came from. That provided a direct link between what astronomers can measure with telescopes and the detailed look that meteoriticists can find in the laboratory. *Hayabusa*'s mission was a great scientific success, despite many troubles along the way. In spite of these trials, or perhaps because of them, it was hugely popular in Japan. Two movies were made about the plucky spacecraft that could.

So far, all the asteroids visited had been of the stony type. Visiting any asteroid is good, but most scientists hanker to get to the carbonaceous types. That's because they think these preserve the best traces of the earliest times in the Solar System. True, Mathilde was a carbonaceous asteroid, but the images from *NEAR Shoemaker* didn't give much away about what it was made of.

But now we are getting our wish. All the public enthusiasm over *Hayabusa* led JAXA to build a second, more ambitious, asteroid probe, *Hayabusa2*.[16] This time the destination is the carbonaceous asteroid Ryugu. *Hayabusa2* arrived in 2018 after a four-year journey, and successfully returned a sample from the kilometer-sized asteroid in 2020. Ryugu has a peculiar top-like shape, round and with a fat equator bulging out. This is a sensible shape for a spinning rubble pile.[17]

OSIRIS-REx is a NASA spacecraft launched toward the asteroid Bennu in 2016. Bennu used to be known by the less romantic name 1999 RQ 36, but the *OSIRIS-REx* team wanted something better. So it joined with the Planetary Society and the LINEAR project (which had discovered the asteroid in the first place and so had the naming rights) to hold a naming competition among schoolchildren. Over 8,000 entries flooded in. The winner was Michael Puzio, then a nine-year-old third-grader from North Carolina, who realized the long arm of the spacecraft would look like a crane's leg as it reached out to the asteroid. That led him to suggest Bennu, the name of a crane god in ancient Egyptian mythology.[18]

OSIRIS-REx arrived at Bennu in 2018. Like Ryugu, Bennu had had its shape modeled, and that too checked out when *OSIRIS-REx* got close enough to take clear photographs. *OSIRIS-REx* carries a powerful suite of instruments. These quickly found that Bennu is rich in water, just as had been hoped.[19] It will bring home a sizable sample of surface rock from Bennu, maybe as much as a kilogram, due to land on Earth in 2023. Bennu was chosen as *OSIRIS-REx*'s target because it is one of the few accessible carbonaceous asteroids we know of. Scientists hope it will tell us about the early years of the Solar System. A kilogram may not seem like much, but the pristine samples from the Tagish Lake meteorites weigh in at less than a kilogram, and there are really no other meteorites so unsullied. Moreover, we don't know where the Tagish Lake meteorite came from. This perspective makes the truly pristine samples from *Hayabusa2* and *OSIRIS-REx* more obviously game changers.

The results from the mission will also help us design better prospecting vehicles in the future. For example, the *OSIRIS-REx* team found that Bennu is throwing rocks off its surface every few days.[20] Fascinating, but also a bit risky for a small spacecraft to get near to. They also found that there is so much rubble strewn around its surface that there are no large clear spaces to land on Bennu.[21] The team found a few places that are just large enough, but it took a precision landing to collect the sample. Given the delay time in signals to and from Earth, that landing was entirely robotic.

Having samples from a couple of asteroids is a great step forward. But telescopic observations show that asteroids have two dozen different types of spectra, and we don't know what all of them indicate regarding what the asteroids are made of. We really want to sample all the different types. That will mean finding ways to get samples back on a budget.

NASA has taken a shine to asteroids. There are two new NASA science missions to asteroids being built, *Lucy* and *Psyche*. Each will go to a new type of asteroid. To be launched in 2021 and 2022, *Lucy* and *Psyche* will each arrive at their targets four years later, in 2025 and 2026. *Lucy* will go to the Jupiter Trojan asteroids and fly by several of them.[22] Trojan asteroids trail Jupiter in its orbit, 60 degrees behind the giant planet at the fifth Lagrange point, a stable location (figure 6). They have counterparts 60 degrees ahead of Jupiter, at the fourth Lagrange point. Naturally these are called Greeks. Most of the time, though, the two groups are lumped together and called Trojans. All of the Trojan and Greek asteroids are named after characters in Homer's *Iliad*, which is about the war between the two. Trojans may be as numerous as Main Belt asteroids but are even farther away and harder to reach. Scientifically they matter because they are probably unchanged from the time of their formation far out in the Solar System, before migrating planets swept them inward to where Jupiter's gravity captured them in these strange orbits.

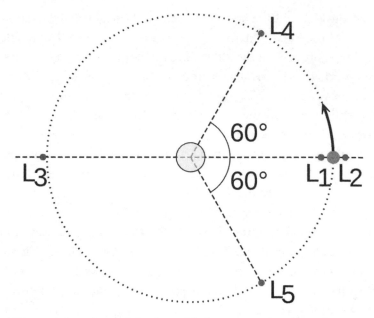

Figure 6: Locations of the five stable Lagrange points for a planet with the Sun at the center. Another body, such as an asteroid or a satellite, can orbit in these points stably. L1 and L2 lie radially inside and outside of the planet's orbit, respectively. L3 lies on the other side of the Sun, opposite the planet. L4 and L5 lie 60 degrees ahead and 60 degrees behind the planet, respectively. Planets, notably Jupiter, gather asteroids at these points, especially at L4 and L5. For Jupiter they are called the Greeks and the Trojans, respectively.

Psyche, the second new NASA asteroid mission, will fly to the Main Belt to visit a new type of world. It will orbit the largest known metallic asteroid, Psyche, which measures about 230 kilometers across.[23] Iron meteorites are common on Earth, but no one has ever visited one of the metallic asteroids of which they are fragments. In surveying a metal asteroid we will see the central core of a planet, or at least a planetesimal, for the first time.

There is a fifth step in the NASA sequence—sending humans to explore. Will we send humans to asteroids? Compared with the logistics involved in sending people to Mars or the other planets, asteroids score surprisingly well. Their low gravity means no dramatic blastoff from their surfaces is needed, substantially cutting down the mass of rocket propellant needed. Still, sending astronauts to a distant asteroid is beyond the ability of today's rockets. New rockets, notably the *Starship* and *Super Heavy* booster combination from SpaceX, will be powerful enough to do so, but we will still have to master life-support systems that can run for a year or two without fresh supplies. Then there's the hazard of the galactic cosmic rays to deal with. Sending a crew to an asteroid doesn't seem likely just yet.

Instead, why not bring the asteroid to the astronauts? It turns out we can, in principle, push a small (house-sized) asteroid from its native orbit to one that's more convenient for us to send people to. We can bring the mountain to Mahomet. That requires a certain chutzpah, a certain boldness of thinking.

I was lucky enough to be involved in the first study of moving an asteroid. The sessions were held at the Keck Institute for Space Studies in 2011 at the Caltech campus in Pasadena. It was one of the most intense, fun, and rewarding weeks I have ever experienced. In a blessed relief from standard procedure, there were no PowerPoint presentations, just a blackboard. No one was allowed to text or read email—under pain of being pelted with soft toys! Instead it was a real brainstorming session, expertly facilitated by John Brophy of JPL.

I started out thinking it was a sort of crazy idea. By the end of the week, though, I was a convert—it seemed doable. As the only regular astronomer in the room, I had the sad duty of pointing out that finding and tracking a house-sized (ten meters in diameter) asteroid would be really, really hard even with the most powerful telescopes we have. An alternative was quickly developed: to pick up a similar-sized boulder from the surface of a larger, easier to track, asteroid. We know that Itokawa, Ryugu, and Bennu are covered in boulders of all sizes. Several radar images of near-Earth asteroids show boulders too. So at least some convenient asteroids have a plentiful supply of good-sized boulders. You would be sure to plan your trip to one of these, as it would be awkward to get out there only to find a smooth plain of solid rock below you.

Amazingly, within a year of this workshop, NASA had taken up the idea and run with it. That's extraordinarily fast for a space agency, and it also got a great response from the various NASA centers, several of which were soon working away at the challenging technologies involved. The result was NASA's Asteroid Redirect Mission, or ARM.[24] ARM was going to bring its boulder back to Earth-Moon space. Bringing asteroids to near-Earth space raises all sorts of questions. How big a rock could be seen as an acceptable risk? If we make a mistake and the boulder crashes to Earth, will the rock break up on reentry? Could we control it as it enters the atmosphere? These are all good questions that could be answered with suitable tests. These concerns meant that a really high and stable orbit around the Moon was chosen for the ARM boulder.

Once in a safe orbit, the ARM boulder could quite easily be visited by humans. The Moon is only a three-day trip away from Earth. Human visits would allow all sorts of testing of how to work with an asteroid. We could also start a search for more meteorite minerals. To do so we'd have to learn to sort through the asteroid with a fine-toothed comb, without disturbing it too much. That's far easier if we have an orbiting laboratory nearby. It's certainly a lot easier than trying to do so with a small spacecraft far from home.

The political winds shifted, however, and ARM was canceled a few years later. The new NASA goal was to go back to the Moon. That's good too. But you have to stick with one thing till you've finished it or you wind up spending billions on half projects and never actually doing anything.

Meteorites are the third way to study asteroids. We don't have to actively bring them to us; we can just let them arrive on their own. They come to us. Meteoriticists are a fairly separate group from the asteroid scientists, with their own special language and techniques. One way that scientists choose what to study is how much they like abstraction. We astronomers are at one extreme. We study strange, exotic objects for our whole career, but we never get to touch the things we study. Yet for us stars, galaxies, and quasars are very real. It comes as a surprise to us that other folks find this a bit too abstract. Geologists don't have that problem. If you kick a rock, you know it is no abstraction. To astronomers, rocks just aren't strange enough to be exciting. Meteoriticists are to asteroid scientists as geologists are to astronomers. The former hold their subjects in their hands.

The problem with meteorites is that we don't know where they come from. Just saying they come from "space" is too vague. We want to know what type of asteroid each one is from. In a few dozen cases whose falls to Earth were recorded there is enough information about their path through the atmosphere to work out their orbit, at least roughly. That does give a hint of where they originated, although it still doesn't connect them to an individual asteroid. And we don't know what they would have looked like if we had been able to see them with telescopes before they landed here. Making the meteorite-asteroid connection means crossing a big gap, and that limits how much meteorites can tell us.

Watching incoming asteroids on their way to becoming meteorites could help. Can we discover small asteroids just before they impact the Earth: the so-called death-plunge asteroids? Meter-sized asteroids impact the Earth about a dozen times a year (though this number is pretty uncertain, being based on only 12 cases). At this size asteroids are big enough that some of their rubble hits the ground as meteorites, but not so big that there's any danger from them.

If we could find them a few days to weeks before impact, we could characterize them telescopically while they are still in space to establish their properties: orbit, spectral types, shapes, and spin rates. Then, when pieces are recovered as meteorites, detailed laboratory analysis can be compared directly with the telescopic spectra. If we had just a few dozen, we could reliably tie telescopic characterizations of a number of common asteroid types to their laboratory compositions.

The first death-plunge asteroid tracked all the way to the ground was 2008 TC3.[25] This was found by Richard Kowalski of the Catalina Sky Survey just 19 hours before impact. That was just enough time to get hundreds of astrometric and photometric observations and a few spectra at telescopes. The astrometry was quickly used to refine the orbit. Luckily, this orbit showed that the asteroid would impact in the northern Sudanese desert. The meteorites from 2008 TC3 were found by a team of students from Khartoum University led by Muawia Shaddad, a professor there. (In one of those small-world coincidences, Muawia and I were students together for our master's degrees at the University of Sussex on the south coast of England. We lost touch but, oddly, both of us went on to study astronomy on a cosmological scale and then—at the same time—we both found our way to asteroids.) The meteorites from 2008 TC3 were designated the Almahata Sitta fall, after a nearby railroad station. "Almahata sitta" means "station six" in Arabic.

2008 TC3 was a lucky case because two-thirds of death-plunge asteroids fall in the oceans, and at least half the remainder will fall in forests and other places where their meteorites cannot be readily recovered. There have been just three death-plunge asteroids spotted since TC3: 2014AA, 2018LA, and 2019MO, but we were less lucky in each case.[26] Still, with greater vigilance, a few per year may be retrievable. Even the ones that don't impact somewhere convenient may tell us something if we have a day or two of warning. That's because as they enter the atmosphere they all will create fireballs as the rocks and volatile materials within them vaporize. If the bright glow of their meteor trails could be studied with fast imaging and spectroscopy from a nearby aircraft, we could learn a great deal about their strength and composition, though not as much as from a meteorite, even without collecting any meteorites from them. Peter Jenniskens of the SETI Institute is already doing this for meteor showers.[27] A new generation of surveys for near-Earth asteroids will have the ability to find a good number of these death-plunge asteroids. Stand by.

6

Fear: How to Handle the Asteroid Threat

When you face a threat, it's best to learn all you can about it. "Know your enemy," wrote the Chinese strategist Sun Tzu 2,500 years ago in his book *The Art of War*.[1] Good advice. What do we need to know when asteroids are our enemy? First, *locate your enemy*: find all the rocks; if we don't know they exist, we can't tell if they are likely to hit us. Next, *track their movements*: work out which ones are on orbits that just might lead them to impacting the Earth. Then, *determine their strengths and weaknesses*: learn how big they are, but also if they are easily broken apart. Once we know which few we should worry about, we can track them with precision, and send spies to the scariest ones to see how best to deflect them.

That step-by-step approach assumes we have time to spare. In reality we may need to do all the steps at once, before a dangerous asteroid moves so far away that we can't make all the careful measurements we'd like. Or before it comes too close for comfort!

Before all else you must find your enemy. The first few asteroids were discovered by Piazzi and the Cosmic Police sweeping the skies looking for "stars" that moved. The same basic plan is used now but, because we have much bigger telescopes and electronic detectors that are far more sensitive than the human eye, we can find much fainter asteroids. Since the Spaceguard program was started in 1992 using NASA funding, the searches have become ever deeper. Starting in earnest in 2004, for more than a decade the more powerful Catalina Sky Survey in Arizona was the leading discoverer of new asteroids, especially the near-Earth asteroids that might be dangerous.[2] Recently the larger Pan-STARRS telescope in Maui, Hawaii, has gained the lead, though Catalina has made upgrades that keep it competitive.[3]

Catalina could track asteroids about a million times fainter than the naked eye can see. Pan-STARRS can find those that are two to three times fainter still. That may not seem like much of an extra step, but there are many more faint asteroids than bright ones, meaning Pan-STARRS roughly doubles the rate at which near-Earth asteroids are found.

All observations of asteroids, about 2 million a month, get sent to a central clearing house, the Minor Planet Center of the International Astronomical Union, which operates as part of the Small Bodies Node in NASA's Planetary Data System.[4] That's quite an institutional mouthful! Surely this is a grand body with a hushed high-tech Situation Room staffed by white-coated geniuses? Well, not quite. The Minor Planet Center (MPC) is pretty much a shoestring operation. It is half a dozen people—all geniuses, of course—down the hallway from me at the Center for Astrophysics | Harvard & Smithsonian. They have a smooth operation, honed over many years, to process the flood of individual observations of asteroids they receive every day. But no Situation Room, and no white coats. T-shirts and jeans are our uniform. (It's the nerdiness of our T-shirts that proudly identifies astronomers.) The MPC catalog lists over 20,000 near-Earth asteroids, and it grows by about 2,000 a year.

That sounds good. There's a problem, though. It will take at least another 40 years to find them all. We don't want to underestimate our enemy! How do we track them all down? Can we do better? We can. There are two complementary projects under way that will find almost all the potentially hazardous asteroids: the mountaintop Vera C. Rubin Observatory's Legacy Survey of Space and Time (LSST), and the space-based Near-Earth Object Surveyor.[5]

LSST will use a large telescope in the dry and mostly cloudless Andes Mountains of northern Chile. The telescope that LSST will use, the Simonyi Survey Telescope, is specially designed to scan the whole of the night sky quickly. For a decade, starting in 2023, it will take pairs of snapshots to find moving objects in the sky, with finding near-Earth asteroids a major goal. The scientists aim to find 80 percent of the near-Earth asteroids bigger than about 500 meters across, and about half of those at 100 meters across. That aim applies to the stony asteroids. But, while normal rocky asteroids reflect about a quarter of the sunlight shining on them, and so appear quite bright, a carbonaceous asteroid typically reflects 10 times less sunlight. So a given sized carbonaceous asteroid has to be about three times closer to Earth to be picked up by one of our telescopes than a rocky one the same size. That means a normal optical telescope can search a much smaller volume of the Solar System for carbonaceous asteroids. Many of them are evading detection. JPL scientist Amy Mainzer (now at the University of Arizona) and her team estimated in 2011 that we have yet to detect about 80 percent of near-Earth asteroids in the 100–300-meter-diameter range, with about 2,000 then known.[6] The carbonaceous asteroids are the ninjas of the Solar System.

The solution, as for prospecting, is to use a telescope that detects infrared light. The sunlight that an asteroid does not reflect is absorbed, warming up the rock. That makes it glow with black body emission at wavelengths in the infrared band. The darker asteroids in visible light will be brighter in infrared light, solving our problem. So why aren't we doing big surveys for moving infrared "stars" in the sky? Because anywhere on Earth is a pretty bad place to put an infrared telescope. The sky is just too bright. That's because the atmosphere has much the same temperature as the asteroids we want to find, so its black body glow is bright at the same wavelengths. But if we leave the atmosphere behind and get into the vacuum of space, the infrared sky becomes very dark. Even better, there aren't as many stars and galaxies cluttering the view as there are in our visible band. (Astronomers will object that their beloved objects of study are not "clutter." That's what they get for calling asteroids *the vermin of the skies!*) So, an infrared telescope in space is the way to go. Then you can see the asteroids move and track their orbits to see if any might hit us.

There was a small telescope launched into space to survey the whole sky in the thermal infrared band. It was called WISE, not only because it was well designed, but also because its full name was the *Wide-field Infrared Survey Explorer.*[7] WISE was a great success and showed that the method works. Asteroids show up clearly in WISE data and, because WISE scanned the sky repeatedly, they could be tracked as they moved, separating them from the fixed stars.

But space is expensive to get to, and the funding for a dedicated killer asteroid hunter satellite has not yet materialized. Amy Mainzer had been leading a team in studies for just such a dedicated mission. Her team's plan is to put a telescope at a special location between the Earth and the Sun a million kilometers from Earth, called the first Lagrange point (or "Sun-Earth-L1" for the cognoscenti; see figure 6). At a Lagrange point the gravitational attraction of the Earth and the Sun are balanced so that a spacecraft can stay put there. Being inside the Earth's orbit, plus some clever baffles to keep out sunlight, would let this telescope sweep the sky for near-Earth asteroids, even those approaching from the daylight side of the Earth, the way the Chelyabinsk asteroid did. In four years of scanning they would find two-thirds of the killer asteroids and many more, less dangerous ones, besides. NASA has decided to go for this mission as part of its Planetary Defense Program. Thomas Zurbuchen, the NASA associate administrator for science, announced that the Near Earth Object Surveillance Mission, or NEO Surveyor, could fly as soon as 2025. That would be good timing, as LSST will be fully operational by then.

Together, LSST and NEO Surveyor will find over 90 percent of near-Earth objects as small as 140 meters across and will tell us how big each one is. By combining their infrared and optical measurement we will measure their albedo, and so get a first idea of what the asteroids are made of. The two surveys will be a boon not only for allaying our fear, but also for satisfying our curiosity and, as we will see, finding an outlet for our greed.

The top priority is to track the asteroids' movements. Presently it takes about three nights of observations to pin down the orbit of an asteroid well enough to say if it is potentially hazardous. That is useful intel. But it is not always enough to let us pick up the asteroid when it next comes around the vicinity of the Earth—its next "apparition," as it is called. And if we can't find the enemy again, most of the point of finding it in the first place is lost. We know a threat is out there, but we can't do anything about it.

Space astronomy helps us out here. The "fixed" stars are not really motionless. Everything moves, under the influence of mutual gravitational attraction. But most stars move very slowly, so for the most part they can be thought of as fixed. Knowing how they move, though, can tell us a great deal about the structure of the Milky Way and about how stars evolve.

So in December 2013 the European Space Agency sent up the *Gaia* satellite.[8] *Gaia* measured the locations of a billion stars with a hundred times more accuracy than any previous survey! By repeating the measurements over five years, *Gaia* revealed the motions of millions of stars out to the edges of our Milky Way Galaxy. The super-accurate *Gaia* moving map of the sky lets us do a better job of tracking asteroids. Any small patch of the sky, such as a large telescope might image, has a dozen or more *Gaia*-tracked stars in it. That lets us use *Gaia*'s star catalog to pinpoint where our asteroids are in the sky much more precisely than we could before. Existing surveys are already switching to the *Gaia* map, and LSST will use it from the get-go. This will help us pick out the ones that might hit the Earth and allow us to find them again. You don't want to lose track of your enemy.

Next we must observe the strengths and weaknesses of any potentially hazardous asteroids. We need to know how dangerous each one really is, and what the best defense against each one may be.

The total mass and the velocity of the asteroid determine how much kinetic energy it has, and so tell us both how big a bang it will make if it hits the Earth, and how hard it will be to divert. This energy is normally expressed as how many million tons (megatons) of the explosive TNT it is equivalent to. That is how nuclear bombs are graded, making for an easy—and scary—comparison.

How do we measure the megatons of an asteroid? The mass of an asteroid is just its volume times its density. If we know its size, we can work out the asteroid's volume easily enough. But sizes are tough to get accurately, if we won't have black body sizes until we have the planned NEO Surveyor mission. For now, those measurements are few and far between. Radar from the ground can give us excellent size measurements and much more, but only for the minority of asteroids that come close to us. All of those will be potentially hazardous, which helps. Losing Arecibo is a big setback. Size alone does not give us the mass of the asteroid accurately because of the large voids within asteroids that seem to be common.

If we could measure the mass of the asteroid directly, all this would be a lot easier. We can use telescopes to do that for asteroids where we see the effects of their gravity. That's the case for those that come in pairs, the binary asteroids. They are about 15 percent of all asteroids. It is also true of asteroids that deflect another asteroid passing close by with their gravity. That's how Psyche was found to have such a high density that it had to be metallic. But these encounters are rare. We could also use the Yarkovsky effect and measure how much the reradiation of the Sun's light changes the asteroid's orbit as more massive asteroids of the same size will move less than lighter ones. But it is a very small effect and, at least for now, challenging to determine. It would take several asteroid apparitions to measure reliably, which would take a decade or more, and we would rather not wait that long.

That exhausts our remote telescopic options.

What we do know for every asteroid is how bright it is in visible light. We can get from a brightness to a size if we know how reflective the surface of the asteroid is. That is the same as the albedo we encountered earlier. Unfortunately, there's a huge range of albedos for asteroids, from about 1 percent up to 60 percent! Unless we know something extra about it, any given asteroid could be dark and big or bright and small. Not knowing which is bad news. If we were off by a factor of four, even, it would mean that we don't know the megatonnage of the asteroid impacting the Earth to within a factor of 60! Is it 1 megaton or 60 megatons? We can't say.

Just knowing the color of the asteroid would tell us whether the asteroid is stony or carbonaceous.[9] That would be a great help because those two types of asteroid account for nearly 90 percent of the near-Earth asteroids, and each type spans only a narrow range of albedo. (Unfortunately, the remaining 10 percent, the X-types, which include the more dangerous metallic ones, spread themselves over the whole range of albedo.) Knowing that an asteroid is rocky or carbonaceous pins down its albedo to a much smaller range, so that its volume—and its megatonnage—can be pinned down too. It's not a perfect solution, but it's a big improvement.

These color measurements have to be accurate to 1 percent, so it takes a pretty large telescope to do it well for the more numerous faint ones. There are many careful corrections that have to be made to the data before you can rely on the results, and only well-trained professional astronomers can really produce numbers that you would bet the planet on. Programs to gather this information are under way. Nick Moskovitz of Lowell Observatory, near Flagstaff, Arizona, has one he calls MANOS, which stands for "mission accessible near-Earth object survey."[10] Mission accessible asteroids are the first ones we would use for mining, as well as for science, and they tend to be potentially hazardous too. Moskovitz's team has put together a raft of telescopes, some owned by Lowell Observatory, some publicly run. They bag about 300 near-Earth asteroids per year. That's a good step forward, but we'll have to scale up these programs if we are to catch up with the 2,000 new near-Earth asteroids discovered each year, and the greater rate soon to come from the Vera C. Rubin Observatory. But with good orbits we can concentrate on the smaller number that are either potentially hazardous or accessible.

Next we send in the spies. Remote observations with telescopes are not enough. To get the asteroids' masses we have to measure their densities. Since many asteroids are as much as 80 percent empty space, we can't get the megatonnage of a potential impact right without a density. The only way to know for sure for any individual asteroid is to go there, taking a radar system with us, or some other device that can see through to its insides. The same space probes could carry other instruments to measure its mass and discover what the asteroid is made of too. That would give us most of what we'd want to know about a dangerous asteroid.

That's not a cheap proposition. Interplanetary travel is challenging. We'll examine those challenges in detail later on, but even using the smallest satellites each one will cost $1 million or, more likely, tens of millions. That makes sending out a hundred spies too costly. That means there's a big payoff in first learning all we can with telescopes to pick out the few that pose the biggest danger.

Once we know all we can gather about the few truly hazardous asteroids out there, how would we go about dealing with them? As with sharks, it's best to punch an asteroid on the nose.[11] Unlike with sharks, we are not trying to scare off the asteroid; instead we are trying to slow it down just a little so that when it crosses Earth's orbit our planet is no longer in the way. (Or we could pull it along on a leash to speed it up a bit. Either way will do.) Slowing a would-be killer asteroid down by just a few millimeters per second will make it miss us a decade later. The method works by "conservation of momentum." Isaac Newton taught us that "every action has an equal and opposite reaction." What this means is that if you throw something one way, you get pushed the opposite way. Rockets work on conservation of momentum too. Momentum is mass multiplied by velocity. The total momentum of a system never changes—it is "conserved." Rockets work by throwing out relatively small amounts of propellant but doing so at very high speeds. That moves the more massive rocket at proportionately smaller speeds in the opposite direction. Multiplying the mass by the velocity of what you throw out tells you how fast your rocket will move, because it will be given *exactly* the same value of mass times velocity. It is smart, then, to eject the propellant as fast as possible. In the case of hitting a killer asteroid, a fast-moving small mass hitting a large one will be stopped, giving all of its momentum to the large one and slowing it down.

But how do you punch an asteroid on the nose? There are four methods: hammers, nukes, tractors, and billiards.

The first method is hitting it with the biggest hammer you can. This hammer is called a kinetic impactor in the trade. No explosives are needed. The problem is that the biggest hammer we can launch is only a few tons and the asteroid is at least thousands of tons. So it's like hitting a brick with a pebble; the brick isn't going to move much. But if the pebble is moving very fast—like a bullet—then the brick will jump.

The European Space Agency wanted to team up with NASA to try a real-life test to deflect an asteroid using a hammer. Why would they bother? Didn't I just explain precisely how much the hammer should slow down the asteroid? Not so fast. We implicitly assumed that the asteroid stays in one piece after we hit it. But in reality, the impact will kick up debris. Just how much debris and which way it goes can add to the momentum change. Asteroid debris that flies backward in the direction the impactor came from will slow down the asteroid even more, as it takes momentum from the asteroid. The problem is that no one knows how big this "multiplier" will be. It could be almost nothing or quite a lot. Knowing the answer makes a difference. It will tell us how big a hammer we need to throw at an asteroid to save the world.

The first version of the project was called the Asteroid Impact & Deflection Assessment mission, or AIDA. The reference to Verdi's opera about the Egyptian princess entombed in a pyramid is appropriate. AIDA has an operatic goal—the first deliberate change of the orbit of a Solar System body—and it does involve a lot of rock, just not a pyramid. But AIDA, like many an opera production, went over budget and was canceled. The United States is continuing with its half of the project, called the Double Asteroid Redirection Test, or DART. DART is the hammer. The ESA mission, AIM (for Asteroid Impact Mission), was to be the audience for the hammer blow.[12]

Because this is just a first test, DART won't be able to move the asteroid much at all. The designers of the mission came up with a clever plan that will let them measure even a tiny change in speed. Instead of hitting a lone asteroid, where it would be hard to tell how much they had moved it, they will hit the miniature moon of the asteroid Didymos. In Greek, Didymos means "twin." The discoverer of Didymos, Joseph Montani, a member of the Spacewatch team, suggested this name once the moon was found a few years later.[13] The little moon, which orbits only just over a kilometer away, didn't have an official name yet, so the mission planners started calling it Didymoon. (Boringly, they were officially calling it Didymos-B for a while. That's no fun. Luckily, they came up with a new name, Dimorphos, meaning "having two forms" in Greek.[14] Not as good as Didymoon, but let's not complain.) Radar measurements show that Didymos is about 800 meters in size—about as far across as the Burj Khalifa tower in Dubai is high—and Dimorphos is 150 meters across—the size of a football stadium.[15] They orbit each other almost exactly twice a day. Didymos gets close to Earth but never crosses our orbit. That makes the experiment safe. It's not a good idea to practice on one that really might hit us if we make a mistake. Best to wait until we have the method nailed down.

Moving a football stadium by hitting it with a hammer is not easy, even in zero gravity. Knowing how much you moved it is pretty hard too. The plan is to split the job into two parts: NASA will home DART in on Dimorphos and hit it hard when it passes near to the Earth in late 2022. ("Near" is still pretty far in space; in this case it means nearly 300 times further away than the Moon.) DART has a mass of half a ton and will be coming in at about 20,000 kilometers per hour, 3,000 times faster than Dimorphos is in its orbit around Didymos. Dimorphos is about 7 million times more massive than that, so conservation of momentum tells us that DART will change the speed of Dimorphos in its orbit by just

1.4 meters per hour. A snail moves faster than that.[16] But that is enough to change Dimorphos's orbital period by 8 minutes, which is not so hard to detect, although the measurements must be carefully done. That's where the spacecraft built by the European Space Agency comes in.

ESA's new plan is still to visit Didymos/Dimorphos, but not in time to witness the impact. Instead its new Hera mission—named after the Greek goddess of marriage—will arrive some four years later. That will miss some of the action, but Hera will still be able to measure the changed orbit of Dimorphos around Didymos accurately to see how effective the DART hammer was. To fill the gap DART will carry a small satellite that will separate from DART before the impact to photograph the event. Hera will carry two small satellites too. They can risk getting much closer to Dimorphos to measure its internal structure and inspect the impact point without endangering the whole mission.

Even if DART/Hera measure a good-sized multiplier, it will still be true that even the biggest hammers we could throw at a killer rock will have a pretty small effect on a stadium-sized asteroid. That means we will need to send it out years ahead of when the asteroid is calculated to hit the Earth if we are to push a truly dangerous asteroid out of the way, so that the small change in its speed adds up to a large change in its arrival time at Earth's orbit. In real life it will also take a while from finding the dangerous asteroid to getting a rocket ready for launch and building the hammer we need. That's why learning how to deal with killer asteroids now by practicing on the likes of Dimorphos has gained a lot more support. If we know what we need, then it can be prepared well in advance and kept in readiness.

Nuclear bombs are another way to punch a killer rock on the nose. A favorite movie of scientists is the 1998 *Armageddon*. But mostly not in a good way. They tend to snigger at the way the Bruce Willis character, a roughneck who rockets into space to divert a giant asteroid from hitting the Earth, deals with the threat. He drills into the asteroid to put a nuclear bomb deep inside so as to blow the whole thing up. I'm not saying if he succeeds or not, or what happens to him—no decades-late spoilers here. At the time, most of us scientists laughed at Bruce because blowing up the asteroid doesn't really help at all: all the pieces are still on the same orbit and will hit the Earth; all Bruce did was make a cloud of smaller rocks that will hit a larger area of the Earth.

But we were wrong to snigger. Newer calculations made by Cathy Plesko and her colleagues at the Los Alamos National Laboratory in New Mexico show that nuclear explosives, "nukes," really are the best way to deflect or disperse killer asteroids.[17] (Does the fact that these calculations come from one of the two U.S. centers for nuclear bomb research make the scientists biased? I really don't think it does. While that may have provided the initial motivation, it didn't alter their calculations. Scientists, like most academics, are strongly allergic to publishing anything that could later destroy their reputations.) Plesko finds that exploding the nuke inside, but not deep inside, the asteroid works much better than exploding it right on the surface. Exploding on the surface would mean that half the energy of the explosion is lost to space. Exploding it in a hole just a few meters deep will mean that much more of the bomb's energy goes into the asteroid. That way much more of the bomb's momentum is transmitted into the rock, and more is kicked out backward. Both effects slow the asteroid down more than a surface explosion. A deep enough crater could be made by sending in a kinetic impactor to hit the asteroid just a millisecond or so before the nuke. The nuke packs a much bigger punch than a kinetic impactor. Then we wouldn't need such a long warning time. Go, Bruce! (Okay, he didn't really need to dig so far down, but that would have spoiled the ending.)

So much for punching a killer asteroid on the nose. What about pulling it on a leash?

Putting an actual lasso around an asteroid and pulling is not—surprisingly—a crazy idea. The Asteroid Redirect Mission could have done just that, though it was eventually canceled. The idea at first was to capture the entire asteroid in a bag, cinch it up, and then push or pull the tightly closed bag plus asteroid all the way back to Earth orbit. You have to have a pretty big bag, but it doesn't have to be super strong. In zero gravity the big rock inside isn't pushing hard on the bag. The rocket would be a scaled-up version of the rockets used to reposition TV transmitting satellites in geostationary orbits around the Earth's equator. These rockets push very gently but can keep pushing for a long time. This is quite different from the more familiar chemical rockets that burn furiously but briefly.

There are plenty of questions about this "strong-ARM" approach to moving an asteroid. For one thing, lots of asteroids, as we saw earlier, are just piles of loose rock with lots of voids bound by a very little gravity. Pushing even gently on one, even when it's in a bag, could make it re-arrange itself drastically, rather like a zero gravity avalanche. That might throw our little rocket off course, or it might break the bag.

A gravity tractor is the most reliable method of pulling on an asteroid. It avoids a lot of potential problems by never touching the asteroid, instead making use of gravity. This is not the tractor beam of science fiction. Two astronauts, Ed Lu and Stanley Love, invented it in 2005.[18] You just put the biggest mass you can to one side of the asteroid and let the gravity of that mass pull on the asteroid. No ropes or bags are needed. It's slow, because gravity is weak. You can change a 200-meter-diameter asteroid's velocity by just 1/5 of a kilometer per hour—about how fast a turtle walks—in one year using a 20-ton spacecraft.[19] A gravity tractor is reliable because gravity is always there, and you don't have to touch the asteroid at all. You do have to keep using rockets to keep you distant from the asteroid, or else it will hit the spacecraft. You must also be sure to keep the tractor's jets firing to the side of the asteroid, or their plumes hitting the asteroid would cancel out the pull from gravity. With enough warning time we can use this method right now. Twenty years would be enough. That's a pretty daunting timescale, however. We'd have to know about the threat at least another five years before that to get the gravity tractor built and launched. Humanity is not so good at dealing with threats that far in the future.

There is a way to speed things up. An "enhanced gravity tractor" could cut the 20 years down to 2.[20] The enhanced gravity tractor spacecraft would pick up a boulder much bigger than itself from the killer asteroid and hold onto it. If the boulder were, say, 10 times more massive than the spacecraft, then the gravity pulling on the asteroid would be 10 times stronger. Pick-ing up a boulder is an idea that ARM was going to test when it became clear that it was too hard to find a small enough asteroid to put in a bag. A test does seem like a smart idea. The catch is that you have to know that there are boulders on your killer asteroid before you can choose to use this method. It would be embarrassing to arrive there, find no good rocks, and have to say, "Sorry, Earth, but we need another 18 years, so we're all gonna die. My bad."

The last technique is space billiards. After the Chelyabinsk event, the Russians were highly motivated to think of plans to stop killer rocks from space. Boris Shustov, the scientific director of the Institute of Astronomy of the Russian Academy of Sciences, told me of this plan at a workshop in Garching, near Munich. His team came up with the idea of using the asteroids against themselves. The problem with kinetic impactors is that they are just not big enough hammers. The Russian group suggests using a spacecraft to hit one of the plentiful house-sized asteroids so that it is deflected to hit the threatening stadium-sized ones. David Dunham of KinetX Aerospace worked with the Russian group and publicized the idea.[21] Good old conservation of momentum means the larger asteroid will be slowed down a lot more than it would be by one of our relatively light-weight spacecraft. Hence "space billiards." There are enough of these small asteroids that one will almost always be in a convenient place to be moved to an intercept orbit with the bad asteroid. Steering a huge mass like this to hit a small target millions of kilometers away will be tricky, but it seems feasible—in principle. The plan does need the smaller asteroid to swing by the Earth to put it on the right track toward the bad asteroid, and to speed it up. We may not want to risk that until we've had more practice at moving small asteroids around the Solar System.

Shustov's team has at least given us a good motivation for finding asteroids as small as 10 meters across. We'll have to be clever to think of a way of doing that, as these small rocks will appear very faint.

Which of these is the best method for saving the Earth? The method you choose will depend on what the threat is and on how much time you have to do something about it. If you have lots of time to act, then the gravity tractor is best because it always works, regardless of what type of asteroid you are dealing with. How the asteroid is spinning or tumbling doesn't matter at all. If the asteroid cooperates a little and has some easily grabbed boulders on it, then the enhanced gravity tractor is even better. But if you have only a few years before impact, then punching the asteroid with kinetic impactors—or, probably better, nukes—is the way to go. If you are facing a really large, fast-moving comet incoming, then playing space billiards is probably the only option.

Some people worry about "dual use"—that asteroids will be co-opted as weapons. If several agencies around the planet can work together to stop an asteroid threat, could one of them alone use the same technology to create a threat? There are many more asteroids that just miss the Earth than ones that hit it. Nudging one of those into an orbit that intersects with the capital city of a perceived enemy could be tempting. That would go double if the perpetrator could blame the destruction on natural causes.

This scenario is unlikely for several reasons. It takes years to change an asteroid's orbit to impact Earth. You would have to have a very long-range plan—decades—to make that a useful weapon. There would also be several years of warning before your enemy got hit; time enough for them to hit you preemptively by more direct means, and time to mount their own deflection campaign. To hide your intent, you'd have to avoid your outgoing rocket being tracked. The Planetary Defense Conference exercises showed that the impact point could move continental distances due to a minor mistake. The precision targeting you would need is much beyond the state of the art, so you've got a lot of development work to do. And then any orbital correction maneuvers you have to make near to Earth would be obvious and would likely mark you out as the perpetrator. All in all, it's far easier to use a bomb on an intercontinental ballistic missile.

7

Greed: Asteroid Prospecting

What does a miner need to know? As Daniel Faber, the original CEO of Deep Space Industries, told me, miners need just 1 bit of information—"Is it ore? [YES/NO] ." Anything else is just supporting information. The word *ore* has a special meaning to miners. Ore is not simply a high concentration of some desirable material. As Mark Sonter, an early pioneer of asteroid mining, points out, "Ore is commercially profitable material."[1] In other words, "ore" is the industry shorthand for "We can make a profit mining this stuff." Daniel's deceptively simple question quickly generates two further questions: "Does it have water?" and "Can we extract it?" and then many more. How much prospecting would be sufficient to flip that bit to YES? Given all the near-Earth asteroids out there, how do you decide which one to visit? There are three big things to learn: the mass of ore available, how easy that ore is to extract, and how easy it is to get to and from the asteroid. That makes it just a question of geology, engineering, celestial mechanics, and rocket science. Sounds easy, right?

Prospecting for ore on Earth involves several stages, and the same incremental approach will likely be used for the asteroids.[2] Terrestrial mining companies begin with preliminary searches of public surveys in order to generate a list of target areas. Then they progress to detailed mapping of targets. Then they must make contact with the ore body, at first by sampling small rock chips. The process then moves on to preliminary drilling to get a better idea of ore size, and after that, to resource evaluation, drilling enough times to sample the ore body to gain confidence in the size of the deposit. Once this is all completed, with success at every stage, a due diligence feasibility study is made that assesses all the factors likely to influence the profit.

The difference with prospecting the asteroids is that it's hard, and expensive, to get to any of them. The early steps that on Earth involve geologists going out to the target site have, instead, to be carried out remotely with telescopes, as detailed in the previous chapter. Only once that option is exhausted will spacecraft go out to the chosen few asteroids to perform detailed mapping from a close orbit. The far more perilous job of touching down on the asteroid to take a rock-chip sample, or of drilling deep within the asteroid, is likely to be a separate third stage. In all of these stages sending a live geologist would be best, but that is not an option—because of both the expense and the health hazard it would be to your live geologist.[3] Instead it will all have to be done robotically. That doesn't mean we can take our time. We should follow the motto of the Blue Origin rocket company: "Gradatim Ferociter," Latin for "Step by Step, Ferociously."[4] We need to begin.

Not every mountain is a gold mine. Since asteroids are mountains flying in space, we should not be surprised if many of them are not really worth all that much. After all, we all know that only a small minority of mountains on Earth are worth mining. A map of the gold mines in Colorado shows that they are almost all in one quite narrow strip of the Rockies (figure 7).[5] This isn't so strange. Ore, by its very nature, is highly concentrated. If gold has to exceed some concentration to make mining profitable, then the distribution has to be spotty: the high points have to be balanced by even more low points. Should we expect that only a few asteroids are ore-bearing too? It's not obvious, as there are fewer geological processes at work for them. However, it does turn out to be the case. That means that aspiring asteroid miners will have to sift through large numbers of duds to find a few good asteroids. That forces miners to first rule out the many asteroids that are clearly not good prospects so as to concentrate their efforts on the remaining better bets, until they close in on a small set of the most promising ones.

We want to know how many asteroids really contain ore. Working out whether a particular mountain—or asteroid—contains enough ore to be

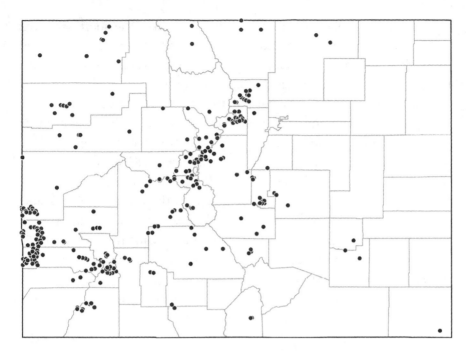

Figure 7: Map of hard-rock mines in Colorado showing how they cluster in a relatively small area.

worth mining has to include consideration of several factors: the total amount of ore, the cost of extracting it and of bringing it to market, and the price it will fetch when sold. So when we prospect for valuable asteroids we need to sieve through the entire asteroid population, not just for those rich in platinum, water, or some other precious good, but for the smaller population of those that would be profitable to mine after taking all of the other constraints into account.

To get a handle on the size of the problem of how many ore-bearing asteroids we can expect, I took Frank Drake's famous approach to the search for extraterrestrial intelligence, or SETI, and shamelessly applied it to asteroid mining. I was hoping that ore-bearing asteroids would be more common than aliens. Drake's goal was to estimate how many civilizations there are in our Galaxy that we could contact. His method was to systematize Fermi's question "Where are they?" by means of an approach that was both simple and, because it was clarifying, brilliant.

He took the number of stars in the Galaxy and asked the question "What fraction of stars have planets orbiting them?" and then "What fraction of those planets lie in the 'habitable zone' where water can be liquid?" That's not yet the answer. He also had to ask, "What fraction of habitable zone planets support life?" and "What fraction of life-bearing planets produce civilizations?" His final question was "What fraction of civilizations produce radio emission we could detect?" Each question reduces the probable number of alien civilizations to a fraction of what was left from the previous question. So if you know all these fractions, then multiplying them together gives the (smaller) fraction of stars in the Galaxy with probable alien civilizations. (He actually worked with rates rather than total numbers, that is, how many stars are born per year and how many civilizations die per year, but it's easier to think about the total number, and that is a more relevant approach to asteroids.)

He summarized the whole approach in what has long been known as the Drake Equation.[6] It looks like this:

$$N = R \times f_p \times n_E \times f_l \times f_i \times f_c \times L$$

That's a long equation, but it shouldn't intimidate you. It just says that the number of detectable civilizations in the Milky Way (N) equals the rate at which stars form (R), times several fractions. They are the fraction of stars with planets (f_p), the number of Earth-like planets around each star (n_E), the fraction with life (f_l), the fraction of those with intelligent life (f_i), and last the fraction of those with good enough technology that we could detect them (f_c). At the end, the whole thing is multiplied by the average lifetime of those civilizations (L).

When Frank Drake wrote his equation on a blackboard at the new National Radio Astronomy Observatory in Green Bank, West Virginia, back in 1961, every single one of these fractions, apart from the rate at which stars form, was utterly unknown and could only be guessed at. For 40 years many people laughed at his equation because of these yawning unknowns. Now, though, thanks to the explosive growth of the field of exoplanet research, and especially thanks to NASA's *Kepler* mission, the first two fractions are pretty well tied down, and the fraction with life is being worked on hard. So far, the answer is promising, assuming that you like the idea of a Milky Way teeming with life. Cixin Liu's science fiction trilogy, beginning with *The Three-Body Problem*, says—spoiler alert—that maybe that would not be such a good thing.

Frank Drake didn't expect to be able to get an answer out of his equation, at least not right away. Instead, the power of his equation is that it helps us think about the problem and points us to the smaller, more tractable problems we have to solve first in order to answer the big question.

The same approach helps us with asteroid mining. We can break down the question of how many ore-bearing asteroids there are into a series of simpler questions that we can answer one by one.[7] We know how many of the known asteroids are big enough that they might contain a billion dollars' worth of materials, $N_{diameter}$. What we want to know is the number of ore-bearing asteroids, N_{ore}. To be ore bearing, these asteroids have to be both rich in the resource we are after—water or precious metals—and reachable with today's rocketry.

There are four steps from $N_{diameter}$ to N_{ore}. First, the fraction of asteroids of the right type (which we'll call f_{type}): that would be carbonaceous ones for water or metallic ones for precious metals. Second, we need the fraction of the ones of the right type that are rich in water or platinum (let's use f_{rich} for them). Third, we need to know what fraction of the ones we have so far are also accessible to our rockets and can return with a large enough payload of ore (call it $f_{accessible}$). Fourth and finally, we care only about the fraction for which the engineering is easy enough to mine that we make a profit ($f_{engineering}$). Multiply all these fractions together and we'll know what fraction of asteroids are ore bearing: whether they are common or rare. The actual number $N_{diameter}$ is just that fraction times the total number of asteroids bigger than our minimum size. That's cumbersome to describe in words. In the concise terminology of mathematics, the equation for asteroid mining looks like this:

$$N_{ore} = f_{type} \times f_{rich} \times f_{accessible} \times f_{engineering} \times N_{diameter}$$

The magazine *New Scientist* dubbed this the "Elvis equation," and who am I to argue with its wisdom?[8] The equation may look scary, but each step is simple, and it closely resembles the Drake Equation. In fact, it is a bit easier because we don't have to worry about how often asteroids are born or about their lifetime. Both are million-year timescales for the near-Earth asteroids, which are the ones that we can hope to mine in the near future. The good news is that with asteroid mining we are in far better shape than Frank Drake was in his equation in 1961. We can already put numbers on every one of the asteroid fractions, at least roughly, for both precious metals and water. Let's evaluate each of the fractions, one by one.

The Right Type of Asteroid, f_{type}

Whether asteroids are of the right type is based on their colors. Those colors give us clues about the type of rock on the surface of an asteroid. We saw that spectroscopy is the tool for this, and it can be done pretty well using mountaintop telescopes, so there's no need to go out to the asteroid to gather this first bit of information. That's why it was worth splitting off as a separate fraction. Spectra can tell us which asteroids are carbonaceous, which are stony, and which might be metallic. If we measure lots of them, then we'll know what fraction are the kind we are after.

It's still a bit tricky, though. For one thing, we only have 1,000 or 2,000 spectra of near-Earth asteroids so far, and only a few dozen of them are metallic asteroids, so the fraction they make up of the whole can't be known super accurately. Just a few errors or omissions could make quite a difference. We also have to recognize that the ones that we have spectra for will tend to be the brighter ones, just because a telescope can't get spectra of very faint ones. That means we tend to get spectra of asteroids that reflect a lot of the sunlight falling onto them (those with high albedo). So we have to correct the raw numbers we see to allow for that bias. This is no small effect. At the same distance, dark asteroids can be as much as 10 times fainter than shiny ones of the same size. That means we have to multiply the apparent number of dark asteroids, which are mostly carbonaceous, by a pretty big number to know how many of this kind really are out there.

Rick Binzel and Joseph Scott Stuart did this carefully at the Massachusetts Institute of Technology back in 2004.[9] They found that, even with these corrections, the dark carbonaceous asteroids made up just 15 percent or so of the near-Earth asteroids, while metallic asteroids made up only about 4 percent. These are their f_{type}. That gives us odds of about 1 in 7 for water and 1 in 25 for precious metals. That's already scarce enough that you will want to know an asteroid's type before heading out to prospect it up close.

Rich in Resources, f_{rich}

Next let's estimate what fraction of the right type of asteroids have a high concentration of the ore we seek, f_{rich}. Telescopes on Earth don't help here. While they can tell if water is likely to be present in an asteroid, for the most part they can't say how much water there is or whether an asteroid is definitively made of metal. Instead we have the meteorites. Laboratory measurements of meteorites can tease out just how much valuable material there is. Mostly the answer is "not much." Getting blood from a stone is famously hard. Getting water from a rock is almost as tough. Eugene Jarose-wich of the Smithsonian measured less than 0.1 percent of water in a quarter of carbonaceous meteorites.[10] At the opposite end, he found about a quarter had over 10 percent of water by weight. He didn't measure all that many, and it would be nice to have measured more carbonaceous meteorites, but to get started we can set f_{rich} for water at one quarter.

Measuring how much platinum and other PGMs are contained in iron meteorites is done with a very different type of spectroscopy. A small sample of the meteorite is irradiated with neutrons by placing it near to a working nuclear reactor and then measuring the gamma rays that it emits as a result. Most of this work was done in the 1960s and 1970s when nuclear reactors were new and could readily be used to irradiate meteorites to reveal the elements within them. Since then this work has fallen out of fashion because the basic questions were answered. What they found back then is that the highest precious-metal richness is nearly 10,000 times that of the poorest! It seems that the metallic cores of planetesimals could not have been at all uniform. Clearly, then, it will be really important to a miner to check the composition of a metal asteroid carefully before going all in on a major mining venture. That will have to be done by prospecting close up to the asteroid. Still, while half of iron meteorites have almost no precious metals, the other half are richer than terrestrial platinum mines. A quarter of them are at least five times richer. So let's set f_{rich} for PGMs to be one quarter.

On a cautionary note, there are only a few measurements of meteorites with the highest concentrations, so even a few new measurements might change that one-quarter value quite a bit. Maybe it's time to go back to the nuclear reactors with more meteorites?

Easy to Get to, $f_{accessible}$

Now we get to the celestial mechanics. How can we best get around the Solar System? The accessible fraction, $f_{accessible}$, doesn't care what the asteroids are made of, only where they roam in space. In space, as we saw, the valuable currency is energy. Our rockets are not nearly powerful enough to dart around the Solar System like the *Millennium Falcon* from *Star Wars*. Instead, we almost always have to use the trajectory that requires the least amount of energy needed to go from one orbit to another.

Doing this right is tricky. Brent Barbee of the NASA Goddard Space Flight Center runs the Near-Earth Object Human Space Flight Accessible Targets Study.[11] That's quite a mouthful, so it is abbreviated to NHATS most of the time (pronounced "gnats"). Barbee uses a powerful super-computer that he borrows from a climate-modeling team to work out trajectories for spacecraft to travel from the Earth to each asteroid and back. He has to calculate a new trajectory for every possible launch date for 20 years ahead, and he has to keep track of all sorts of details that matter when planning a space expedition: When can you leave? How long does the trip out take? How powerful a rocket do you need? How fast can you come back into the atmosphere? (Preferably not too fast!) Are there ways to cut the journey short if need be?

That's a demanding list. It would be handy if we could heavily prune the list before digging into all the details. A handy number for doing so is delta-v. Delta-v is the total change in velocity a spacecraft must achieve to make the trip to the asteroid, usually starting from a low Earth orbit, or LEO. It's measured like the speed of your car in a distance per time step. Kilometers per hour would be a really large number—18,000 kilometers per hour or more—so we normally use kilometers per second, which turns out to be a handy small number from about 5 to 10. Delta-v is a measure of the energy per ton you need to get the spacecraft to the asteroid from a starting orbit. (It's the square root of twice the energy per ton, to be precise.) Usually people assume LEO as their starting point.

The value of delta-v for a given near-Earth asteroid is absolutely crucial to knowing if we can get to it with any useful cargo. This is because of the "Rocket Equation." Rockets have to accelerate not only their cargo, but also the fuel they burn to change orbits. The faster you want to go, or the more you want to carry, the bigger the tank of propellant you need. This leads to rapidly diminishing returns. The Rocket Equation calculates the penalty a rocket has to pay for carrying its propellant with it, instead of using the surrounding air the way an airplane does. The Russian Konstantin Tsiolkovsky was the first to derive the Rocket Equation in the context of space travel. In his 1903 book he presented his Rocket Equation to show that a rocket could reach Earth orbit, something no one had thought about before. The Rocket Equation is merciless. Small increases in delta-v lead to large drops in the payload mass that can be delivered to the asteroid. The starting mass of a rocket actually grows exponentially as the final mass increases. Launching a satellite of double the mass means starting with a rocket that is over seven times heavier. As a result, adding delta-v exponentially decreases how much payload can reach the orbit of our chosen asteroid, whether it be experiments, explorers, or mining equipment. All in all, low delta-v is a very good thing for aspiring asteroid miners.

A first estimate of the outbound delta-v can be calculated fairly simply with some formulas developed by the German engineer Walter Hohmann in 1925. There were no computers back then, so he had to find a simplified way to calculate the energy needed. Also, rockets were still a really new idea. Robert Goddard (after whom the NASA Space Flight Center in Maryland is named) didn't fly the first rocket powered by liquid fuel until a year later. What Hohmann worked out are the orbits that use the least possible energy to go between a pair of Solar System bodies. Delta-v from the surface of the Earth to low Earth orbit needs about 10 kilometers per second. From LEO to escaping the Earth altogether is just another 3.2 kilometers per second. Matching velocities when you arrive at the target body takes another delta-v, with the value depending on what the body's orbit is. These trajectories are now known as Hohmann transfer orbits.

Hohmann's idea was developed for asteroids by Eleanor Helin and the same Eugene Shoemaker of Barringer Crater fame. Helin led the Near Earth Asteroid Tracking program at JPL. Lance Benner, also of JPL, uses the Shoemaker-Helin method to keep a handy table of delta-v values for all the known near-Earth asteroids, in case you want to check on your favorite asteroid.[12] They have delta-v values from about 3 kilometers per second up to about 12 kilometers per second, with most of them having 6–7 kilometers per second. All the spacecraft sent to near-Earth asteroids so far have delta-v of 5 kilometers per second or less. Disappointingly, only 2.5 percent of near-Earth asteroids have such a low delta-v.

Reducing the outbound delta-v from the average value to 4 kilometers per second doesn't sound like a big difference. Yet it doubles, and can even quadruple, the payload that can be delivered. This sensitivity makes the few asteroids with ultra-low delta-vs of great importance. Many whole classes of near-Earth asteroids are ruled out. For now, the asteroids in the Main Belt are inaccessible to miners, as my student Anthony Taylor confirmed.[13] They need much too large a delta-v to let us transport heavy mining gear to them using today's rockets. We'll just have to stick with the near-Earth asteroids at first. Even among near-Earth asteroids those on highly inclined orbits to the Earth's are of no interest for early mining operations, or for human exploration, as they have a high delta-v. The few percent of low delta-v asteroids tend to have orbits very like the Earth's. That also means they have a greater chance of hitting us, giving them a two-faced interest to us—greed and fear combined.

What threshold should we then set for $f_{accessible}$? The most ambitious mission to return an asteroid sample to Earth is *OSIRIS-REx* and that plans to return just a few hundred grams of rocks from the surface of the asteroid Bennu. Bennu has a delta-v of 5.1 kilometers per second. Since in the best case we miners will need to process an entire 1,000-ton asteroid to bring back 100 tons of water to sell, we surely won't be sending our big tough mining robot spacecraft to any near-Earth asteroid with a larger delta-v than Bennu. That gives us access to just 2.5 percent of known near-Earth asteroids; $f_{accessible}$ is 2.5 percent.

Easy to Engineer, $f_{engineering}$

The last term is $f_{engineering}$. This fraction lumps together all the tricky engineering problems associated with actually extracting the valuable resource and bringing it home. This is a huge and complex field that is only just starting to be investigated.

I am no engineer. That's despite my entire family being engineers—my father, my brother, and my mother were all in the trade in my hometown of Birmingham, United Kingdom. I was the family rebel, wandering into the stranger pastures of physics. When I announced that I wanted to be an astronomer my father said, "There can't be more than one job a year in that!" "But I only want one," I replied impishly and innocently. He was right; I was reckless; I just didn't realize it. Luckily it all worked out fine. I think he ended up being rather proud of his wayward son, even if he thought astronomy had no practical value. Too late, I started down this road to "astronomical engineering" after he died at the age of 94. That's a pity. He'd have been really interested and helpful.

So my musings on this topic are pretty basic. I expect that others more qualified will break $f_{engineering}$ down into several separate factors. As a way of getting started, let's just say that every asteroid is mineable, so $f_{engineering}$ = 1. That means that our answers will err on the optimistic side.

The Bottom Line, N_{ore}

What do we have for the bottom line? How many suitable asteroids for mining are there in total? We have all the fractions in the Elvis Equation, and we can multiply them all together. What do we find? For precious metals, f_{ore} = 0.025 percent. That's just 1 in 4,000 near-Earth asteroids. For water, f_{ore} = 0.09 percent, or 1 in 1,100. Clearly, finding a good asteroid to mine is not going to be a cakewalk. It will need a lot of hard prospecting work.

Those numbers are pretty comparable with mines on Earth. Although the success rate for terrestrial mines that are begun is about 1 in 5, that goes down closer to 1 in 1,000 of the potential mine sites that are explored. The world spends about $10 billion a year on exploration of possible mine sites.[14]

That only tells us the fraction. What about the total number of ore-bearing asteroids, N_{ore}? To get that we first have to choose a threshold size. To do this we have to decide how much we want the ore to be worth when we get to sell it. Given the cost of doing business in space at present, let's take $1 billion as a threshold. Anything less just isn't exciting enough for a venture capitalist to get the project going. Following the lead of super-model Linda Evangelista, venture capitalists don't wake up for less than $1 billion.[15]

As for any type of ore, the total value in an asteroid is simply the number of tons of ore in the asteroid times its price per ton. And the mass of ore is just its concentration, the amount per ton, times the mass of the whole asteroid.

Let's start with water. The price that asteroid water could command is no more than the $10 million per ton to equal the historical cost of bringing water from Earth to a low Earth orbit, or $20 million per ton to a high orbit. That "from Earth" cost is going down, but let's use the old price as an optimistic starting point. We want to mine only those asteroids that are at least 10 percent water. A ton of unprocessed raw rock from a water-rich asteroid is then worth about $1 million to $2 million. That means that to add up to $1 billion worth we'll need to start with an asteroid of 1,000 or 2,000 tons. How big an asteroid do we need to have that much raw material? The density of solid rock is about 3 tons per cubic meter. But, as we saw, asteroids are not solid rock, and may be as much as 2/3 empty space, bringing us down to about 1 ton per cubic meter. So, we need 1,000 or 2,000 cubic meters for $1 billion. Fortunately, that is not so much. It corresponds to an asteroid just 9 meters across. Pretty much any asteroid that we can actually find with our telescopes will be bigger than that. Right now, we know of only about 20,000 near-Earth asteroids, so we expect only about 18 to be ore bearing, but all of them, most likely, will contain much more than $1 billion worth of water.

Precious metals are more discouraging. The price of platinum and palladium on the market is similar to that of water in a high orbit, and the price is not coming down. That's good. The bad news is that even the most precious metal–rich metallic asteroid has only 30 grams or so of them per ton. That means we have to mine 10,000 times more raw asteroid rock to yield a ton of platinum group metals than we do to extract a ton of water. Any billion-dollar asteroid then has to be over 20 times bigger, or nearly 200 meters across. The Elvis Equation predicted that there were only going to be about 10 of these in the catalogs we had in 2015.

These numbers for both water and precious metals could be seen as disappointing but, looking at it another way, even before we have truly started, we know there are billions of dollars in ore among the asteroids.

Is there some way to have a bigger inventory of ore-bearing asteroids? We can use our equation to see which of the fractions might be improved.

You might say, for example, "Ah, but what about all those undiscovered near-Earth asteroids you told me about? Maybe there are lots of low delta-v asteroids we just haven't found yet!" The number $N_{diameter}$ that we used was not the total number out there, but just the ones we know about now. Fortunately, our surveys are woefully incomplete. The WISE mission has allowed a good estimate of how many near-Earth asteroids there are altogether, at least for those bigger than 100 meters across, which is the size we need for precious metals, and plenty big enough for water. The answer? Twenty thousand, more or less.[16] So far we know of about half of them, so a complete survey would get us up to about 20 precious-metal ore-bearing asteroids. If we could find all the asteroids down to 50 meters in size that would give us another 30 or so, which would help quite a bit. That we are looking for water in smaller asteroids helps greatly because there are lots of them. There are estimated to be a total of around 100 million near-Earth asteroids bigger than 10 meters across. If 1 in 1,100 asteroids is rich in water, then the total number of water/ore-bearing asteroids is 95,000. So water looks promising once we have that inventory.

What about the fraction of asteroids of the right type to be ore-bearing, f_{type}? That could go up if we could use the subtler features of asteroid spectra to infer their composition. There are 24 sub-types of asteroids based on these subtle signatures, but our job is complicated by the fact that we don't know what all these sub-types mean for their detailed composition.[17] As a result, their resources and their value are unclear. The search for small incoming "death-plunge" asteroids that will create meteorites could help here. There has been progress. The ATLAS network of telescopes is set up to do this. John Tonry of the University of Hawaii invented his Asteroid Terrestrial-Impact Last Alert System to give warnings of asteroids about to hit the Earth.[18] Every night, he and his colleagues scan the whole sky that can be seen from their telescopes on the Hawaiian mountaintops of Haleakala and Mauna Loa. Having multiple sites means that they hardly ever have clouds on all of them, so they miss very few near-Earth objects due to bad weather. The main goal of the ATLAS team is to give three weeks' notice for the impact of a 140-meter "region killer." The smaller and more common 50-meter-diameter "city killers" will be fainter, so they will only be picked up a week before impact. That may be enough time to evacuate the impacted area. It certainly is better than being taken completely by surprise. Still smaller asteroids still will give only a day or two of warning. For our prospecting purposes that could still be enough time to get some spectra of small incoming asteroids, and then to rush to their impact sites to pick up samples or use airplanes to get spectra of their debris as they burn up. Comparing the lab or aircraft measurements with the spectra from when the rock was still in space will fill in the blanks on what resources these other asteroid types contain. It will take some time, though, to get enough samples to make reliable predictions.

Can we improve on the fraction that is rich in a resource, f_{rich}? Maybe. Only a handful of examples of resource-rich meteorites are known, so our estimates could be off quite a bit if we happen to have looked at particularly resource-rich or resource-poor examples. More meteorites could change the answer either way. Such uncertainty is not good for business planning. Perhaps we need an equivalent of the Human Genome Project that would accurately survey all of the 50,000 meteorites in the world's collections. Or we could be more selective. Many of them fell to Earth long ago and so may be chemically altered. That would badly affect water estimates, but would not change precious-metal estimates, as these elements don't react chemically all that much. The cases where we have multiple meteorites from the same asteroid are useful, as they will tell us how uniform they are and so how much we can rely on one sample for an accurate assay.

It would be better if we had many more meteorites from fresh falls. Then we could look at meteorites that have had little time to be altered chemically. The ever-inventive Peter Jenniskens has pioneered a new way of finding fresh falls by using weather radar. It turns out that meteorite falls show up with a distinctive signal in the radar data. He had a startling early success at Sutter's Mill (which is conveniently close to his laboratory and, coincidentally, where the California gold rush started).[19] His team was able to collect the meteorites before it rained so they were almost pristine. Unfortunately, his success alerted the weather radar folks to this "noise" in their data and they are now carefully screening it out of their maps. Surely, we can get them to send it to the scientists instead of tossing it out completely.

The biggest improvement we can hope for is in the fraction of accessible near-Earth asteroids, $f_{accessible}$. Just increasing the highest delta-v we can reach from 5 kilometers per second to 6 kilometers per second gives us about 20 times more asteroids. This would get us up to 200 ore-bearing metallic asteroids. Even without more powerful rockets, a "slingshot" trajectory around the Moon could buy the extra kilometer per second that we need, at least for some near-Earth asteroids. Better rockets are on the way. No new physics, no warp drive is needed. We already have more powerful and cheaper regular combustion-powered rockets coming from SpaceX, and soon from both Blue Origin and United Launch Alliance (ULA) too. If we had fuel depots in low Earth orbit, they could refuel, as at least two of them are planning on. Refueling in space will, of course, be a lot easier once asteroid mining actually gets going and can supply the fuel. We may soon be able to send tons of mining equipment to half of all near-Earth objects instead of to just a few percent.

Putting all the factors together we could well go from just 10 precious-metal ore-bearing asteroids in our catalogs in 2013, when I first estimated the number, up to a couple of thousand within a decade, and maybe many more for water. Each one of them would be worth $1 billion or more. That's beginning to sound like real money. Maybe talking of space trillionaires is about to become plausible enough to wake up those venture capitalists? Once you can say for sure "There's gold in them thar hills—but not in these others," you are on your way to wealth, whether the mountain is on Earth or traveling in space.[20]

Now how do we choose the few most resource-rich asteroids? This matters a lot. If we choose poorly, we'll be out of business. Silver miners didn't dig in just any mountain in the Rockies; they prospected many until they found a good one, such as silver rich Aspen Mountain. Once we have used astronomy to find the right color asteroids, we will need to sample those asteroids to find out which are rich in ore. That means going there, which is expensive. How will we actually go about prospecting for those rare ore-bearing asteroids?

Start with astronomy. To execute this plan there will need to be a change in how astronomy is done. Astronomy is often called the second oldest profession (after accountancy), but it is about to spawn one of the newest professions: industrial astronomy.[21] Astronomers once were useful. They supplied astronomical almanacs for navigation, and long before that they kept the calendar for farming. The invention of spectroscopy changed all

that. The past two centuries have been pure pleasure for astronomers—
understanding the entirety of the universe with no thought for practical
application. The result is magnificent. We now know that the universe
we inhabit is vastly bigger, older, and more complex than anyone had
imagined. There is much more left to discover along the same lines of pure
research.

We have, perhaps, gotten too used to that purity. Now, if asteroid mining is to prosper, a new industrial astronomy is needed. Industrial astronomers will deploy their tools and skills to select the few promising
asteroids to mine out of the many thousands of possible ones we already
know about. Just as geologists are mostly employed by the extractive industries, many astronomers may soon find themselves working outside of
academia.

The standard astronomical techniques of astrometry, photometry, and
spectroscopy are all special skills and all of them are needed for the remote
prospecting of asteroids. To make the accurate observations that mining
companies will need requires professional PhD astronomers. To obtain
the high-quality observations will require large telescopes, and large volumes of observations. The mining companies will need thousands, and
perhaps tens of thousands, of asteroids with high signal-to-noise well-
calibrated spectra, light curves, and accurate orbits. Having a large inventory will give the miners hundreds of good ore-bearing candidate asteroids,
so they can be assured that they will be able to launch a mission at any
time to go to one of them, instead of waiting for months for a good one
to come along. That will cut costs and reassure investors that asteroid
mining is not a one-shot deal.

Some of the large telescopes of today will be retiring from their standard
programs of observing the universe because a new generation of even larger
giant telescopes is coming online in the 2020s. These retired telescopes
are still powerful, but they will need new uses. If even one could be used
for industrial astronomy it would revolutionize our knowledge of space
resources. We may even be able to use somewhat smaller telescopes, at
least for astrometry, if a computer-intensive method of adding many images together of moving objects can be made to work well. This approach,
invented by Mike Shao of JPL, is called "synthetic tracking" and is being
pushed by the B612 Foundation, among others.[22] If space missions get

cheaper, then an industrial telescope working in space would be another great tool to deploy to search for water, as there is a telltale signature of water in the infrared that lies at just the wrong wavelength to detect easily from the ground.

All of this implies an industrial scale of data production that many astronomers will find alien. Not all of them though. There are several large astronomical survey projects that deal with millions of spectra and image billions of stars and galaxies. The astronomers working on these surveys may not know much about asteroids, but they have the skills that companies will need to carry out their large telescope-based projects. Ironically, the new activity in LEO that could benefit from asteroid resources is also threatening the ability of mountaintop observatories to find those resources. The satellite "megaconstellations" of thousands of communications satellites will degrade the images of faint asteroids. This could be a substantial hit.[23]

It may well be more important to the asteroid-mining companies to choose Big Data astronomers rather than astronomers who know all about asteroid types. Our theoretical understanding of asteroids and meteorites is also too undeveloped to be a reliable prospecting guide. Why, for example, does the richness of metal meteorites vary so much? Is the fraction predictable? Pure research into Solar System origins and history is needed more than ever so as to best leverage the data. For this companies will need real specialists in asteroid science.

As industrial astronomy grows and develops standard techniques, it is likely that specialized training at the master's degree level could meet most of the skills needs of asteroid-mining companies. Only a few industrial astronomers will then need the training that a multiyear PhD program provides. This new need is an opportunity for forward-thinking universities. Already the Colorado School of Mines in Golden has a Space Resources Program that awards master's degrees and PhDs.[24] The Luleå University of Technology in Kiruna, Sweden, has started an Onboard Space Systems team that teaches asteroid engineering. The Florida Space Institute at the University of Central Florida is also ramping up its offerings in space resources. These courses deal with more than the astronomy of asteroid mining, including the extraction of the ore. We'll get to that in the next section. The bigger universities have not yet taken up the challenge. I expect some of them will soon.

One change that industrial astronomers will have to get used to is that their work will be the intellectual property (IP) of the companies they work for. Geologists who work for oil companies are used to this. But the new industrial astronomers coming from academia will have to accept that they won't be presenting their latest results at conferences as they are used to doing. This will be a quite a culture shock for them.

Astronomy can supply accurate orbits, sizes, spin rate, and an idea of the surface minerals on an asteroid, and then its job is done. To go further requires an up close and personal visit to the asteroids that look promising. We can call this proximity prospecting.

One prospecting spacecraft that could fly by half a dozen asteroids in a few years would be cost effective, gathering multiple asteroids' details for the price of one mission. It would probably pass hundreds or thousands of kilometers from each asteroid, though. A camera would give the asteroid's size and shape, and which way it is spinning. If the surface is not all the same composition, the camera could map out the minerals on the asteroid's surface. Radar could measure how far away the asteroid is and learn how rough and how reflective of radio waves its surface is. That's a lot, but it would not tell us how rich the asteroid is in resources.

For more detailed information we need to have our prospecting spacecraft stop at an asteroid for a few months. Going into orbit around a small asteroid is not feasible as its gravity is too weak. But co-orbiting with the asteroid is feasible, as the *Hayabusa 1* and *2* and *OSIRIS-REx* missions have proved. With an extended stay, many more tools become useful. Some valuable radiation from the asteroid can be quite weak. With enough time exposed to these signals, though, they can provide telling information. It will be worth the wait if the asteroid seems promising.

One of my favorite tools uses the X-rays generated from the hot corona around the Sun. (X-rays are just light with a very short wavelength, about 1,000 times shorter than visible light.) Those X-rays that fall onto the asteroid make individual elements glow at particular wavelengths, just as Kirchhoff found for gases in normal optical light, and an X-ray detector can make images of the asteroid in each of these wavelengths, giving us a map of the elemental composition of the surface. For a long time asteroids were assumed to be uniform in composition, but at least some are quite varied. Bennu is an example. Its surface is covered with dark-colored boulders strewn on lighter-colored background rock (see figure 1). The small student-built X-ray mapper REXIS on board *OSIRIS-REx* was designed to give our first answers.[25] It could be the first of many.

Approaching to within a kilometer or so of the asteroid would allow the spacecraft to (briefly) free-fall toward the asteroid and so measure how strong its gravity is from how rapidly the spacecraft accelerates toward it. To detect this acceleration we may make use of the accurate distance measurements provided by lidar. Lidar uses laser pulses and times their return to repeatedly work out how far away the object being illuminated is. (Lidar means "light detection and ranging," just as radar comes from "radio detection and ranging.") *OSIRIS-REx* has a lidar instrument, OLA (*OSIRIS-REx* laser altimeter), provided by the Canadian Space Agency, for just this purpose.[26] The inner structure of the asteroid could also be probed with ground-penetrating low-frequency (long-wavelength) radar.

If a still closer approach could be made, while still not making contact with the surface, then it may be possible to fire a laser at a point of the surface to boil off (ablate is the technical term) some surface material, and then use a spectrograph to see what the resulting vapor is made of. This technique, helpfully called laser ablation spectroscopy, would be great to have, but may need too much power in the laser for a small mission.

The last, somewhat wild, suggestion, made by Branden Allen at Harvard, is to use muon imaging. Muons are subatomic particles that are made by cosmic rays interacting with the atmosphere. These particles go through large amounts of solid rock, but they are not like their more ghostly cousins, the neutrinos. Those particles can traverse the whole Earth without being absorbed. Muons instead do get absorbed noticeably in a few meters of rock. This property has allowed muons to be used to see the hollow lava vents within volcanoes and hidden passages within the pyramids.[27] It is just possible that the muons generated when cosmic rays hit their surfaces could be used to see inside asteroids too, exposing the voids within them to contribute to an accurate assay of the asteroid.

The final challenge will be landing on the asteroid and getting a core sample to show that the surface composition is not tricking us. That requires staying on the surface for much longer than the brief encounters that *Hayabusa2* and *OSIRIS-REx* had. Once in place, how do you then drill a sample deep into an asteroid to prove your ore body? It's not simple.

The "heat flow and physical properties package" instrument (known as the "mole") on the *InSight* mission to Mars had a lot of difficulty digging into the surface, even with the help of one-third of Earth's gravity.[28] Having essentially no gravity, asteroids will pose a bigger challenge.

Who will the interplanetary prospectors be? Robots. The '49ers, the first prospectors of the California gold rush, traveled light. Asteroid prospectors will have to do the same. Even after industrial astronomers locate the best candidates, mining companies will probably still have to prospect about 10 promising asteroids up close to have a 90 percent chance that the asteroid really is ore-bearing. That means that their asteroid-prospecting spacecraft will have to come in flotillas. Because we'll need a lot of them to build up an inventory of valuable asteroids, interplanetary prospecting spacecraft will need to be far cheaper than the hundreds of millions of dollar costs of NASA's big planetary missions. The implacable Rocket Equation means they must be lightweight. All of this rules out people doing the on-site prospecting. The *Arkyd* spacecraft developed for prospecting by Planetary Resources followed this imperative; they were shoebox-sized and easily lifted by a single person.[29]

There are three big challenges to making a small but capable interplanetary spacecraft: propulsion, communications, and navigation. Propulsion means having rockets good enough to get to the asteroid. Communications means being able to send back enough information. Navigation means being able to find your way to the asteroid. Traditional planetary spacecraft have relied on bulky rockets with lots of fuel, and on the giant radio dishes of NASA's Deep Space Network and its equivalents for both locating the spacecraft in deep space and communicating its data back home.[30] None of these techniques will work for the large numbers of small spacecraft that will make up the commercial prospector fleets. Luckily there are promising technologies developing rapidly that can solve each one of these problems: solar electric propulsion, optical communications, and X-ray navigation.

The propulsion problem is that of packing enough power into a rocket and its fuel while keeping it lightweight. That is hard. In fact, it really isn't possible with conventional chemical rockets. Why can't we just give the prospector spacecraft a good shove and let it coast all the way to the asteroid? In space there's nothing to slow it down, after all. It's true that the launching rocket can get the prospector spacecraft en route to the asteroid, but if it is to do more than take a few snapshots as it swings by, then it has to decelerate to get into the same orbit as the asteroid. That needs quite a bit of delta-v. The way out is to recall that rockets work on momentum. It's the product of mass and velocity that matter—the momentum. You can get the same delta-v by throwing a heavy thing out slowly or by throwing something a tenth of the mass 10 times faster. Burning rocket fuel uses chemical reactions to heat up gas and send it out of the rocket nozzle. Chemical rockets can produce a lot of thrust, but there's a limit to how fast that exhaust gas can move.

An alternative is ion engines.[31] Ion engines send out their propellant 10 times faster than any chemical rocket. The trick is to create a gas that is electrically charged and then put such a high voltage on it that the gas accelerates to these high speeds. Now you need only a tenth as much propellant to generate the same momentum. That makes the whole spacecraft lighter and the amount of ore you can carry back goes up. Ion engines' thrust is quite feeble, but they can keep on pushing for a long, long time, and that adds up to a large change in delta-v. Electrically charged gas is ionized, and ionized atoms are called ions; that's why this type of rocket is called an ion engine. Another name for it is electric propulsion, because of the high electric voltage needed, and this is the term that is most used now. The electricity that provides the high voltage is generated from solar cells, so another term, "solar electric propulsion," or SEP, has become standard. In principle the electricity could be supplied in other ways, with nuclear power being a favorite. That would be nuclear electric propulsion (NEP). While it would be useful in the outer Solar System (Jupiter and beyond, where the Sun's rays are weak), there are both technical—it is hard to make a tiny nuclear reactor—and obvious safety issues to NEP. Instead, SEP has come along fast.

Early problems with the exhaust stream corroding the electrodes that supply the high voltage have been overcome with the so-called Hall thruster design, which creates a charged plasma to attract the ions so there is no solid cathode to erode. Most communications satellites now use SEP to reach and maintain their geostationary orbits. The NASA *Dawn* spacecraft was able to visit two of the largest Main Belt asteroids, Vesta and Ceres, in 2011 and 2015, thanks to its use of SEP. Substantially more powerful ion engines were already developed for the (now canceled) Asteroid Redirect Mission and will be used on DART. The stage is set for SEP rockets to power prospector spacecraft to the near-Earth asteroids.

Communications matter, as we'll need to send a lot more than Daniel Faber's 1 bit, thumbs up/thumbs down ideal. We will want to evaluate the evidence ourselves. Radio is the obvious way to communicate the data from the cameras and other instruments on a prospector spacecraft back to Earth. And that is what has always been done for interplanetary space-craft. But for a small spacecraft, over the distances of interplanetary flight (1 AU or more), this requires a dauntingly large antenna and a lot of power. A promising answer is to use lasers. A basic fact about radio is that higher frequencies can carry more information. The light we can see is just the same electromagnetic radiation as radio, but at 100,000 times the frequency. Visible light can then carry a lot of information for the same power, or the same information but at much reduced power. A laser also provides just the bright, tight beam we need to avoid light spilling far beyond the Earth and wasting power.

NASA has tested this technique of optical communications (or optical comm, to those in the know) from the Moon to the Earth on its Lunar Atmosphere and Dust Environment Explorer, or LADEE, mission. The receiver for the signal is a normal modest-sized astronomical telescope one to two meters in diameter. It is equipped with especially fast detectors that can follow the very rapid modulation of the laser's intensity. That modulation encodes all the information being sent back. As clouds block the laser light, it's wise to have several at different sites so that the weather cooperates on at least one of them. Astronomical observatories are already in place at some of the most cloudless places on Earth, so they may develop a sideline in being ground stations for space prospectors. Optical comm is promising enough that there are start-ups, such as Analytical Space in Cambridge, Massachusetts, hoping to build super-fast and cheap optical comm networks to provide ever higher bandwidth internet globally. So this technology is poised to solve the prospectors' data problem.

The final problem is navigation: knowing where you are. NASA's Deep Space Network is used by all the agency's present interplanetary missions.

But it can support only a few spacecraft at a time, and it costs a lot to operate, over $1 million per year per mission.[32] That puts it out of range for a commercial company wanting to operate a dozen or more small prospecting spacecraft simultaneously. The ideal would be for each space-craft to navigate itself so no expensive mission control is needed. An autonomous driving spacecraft ought to be a lot easier to set up than an autonomous driving car, what with no weather, signs, pedestrians, or other traffic to worry about. The problem is not the computing but getting the information you need to make the calculation of your position. Surely, like sailors on a ship, you could use the stars to steer by? But when you are floating not on an ocean on a nearly spherical and steadily rotating Earth but out in space, sighting on the stars tells you only which way you are pointing, not where you are. The stars are so distant that they have effectively the same positions when seen from anywhere in the Solar System. You can use the positions of the planets and the Sun, which helps greatly. Sighting on two planets—Jupiter and the Earth, say—plus the Sun is enough to locate a spacecraft anywhere in the ecliptic plane—the plane of the Earth's orbit around the Sun. Repeating the measurements at intervals also tells you how fast you are moving in that plane. That needs only a small visible light telescope on board to measure their positions. The near-Earth asteroids, though, don't stick to the ecliptic plane, but have orbits inclined to that plane. We would likely start by mining the ones that orbit near to the ecliptic plane, as they mostly have lower delta-v. Even then we need to know how much above or below the plane our prospector is, as the accuracy we need to successfully reach a tiny asteroid out in space is really high. Even an inclination of 1 degree reaches over 2.5 million kilometers away from the ecliptic plane.

A novel technique called X-ray navigation, or XNAV, could solve the deep space navigation problem and is having its first trial on the International Space Station. It uses discoveries from my original field of research, X-ray astronomy. It turns out that there are accurate clocks in the sky that send out precisely regularly timed pulses of X-rays. If we use an X-ray telescope to collect them, we can use the phase of these pulses to get our location and use their Doppler shift, which slightly changes their rate, to get our speed and direction of motion.[33] It takes at least three of these

"millisecond pulsars," and preferably four to provide a check on our calculations, to get this information. If everything works perfectly, having an
X-ray telescope on our prospector can pin down its location to a few tens
of kilometers. The NICER experiment on the ISS has shown that the
method works well, giving positions to about 10 kilometers.[34] (NICER is
short for "neutron star interior composition explorer," which is its other
job.) Compact, lightweight X-ray telescopes that would fit on small prospector spacecraft are also in development. As X-ray telescopes are also good
for prospecting once the spacecraft arrives at the asteroid, they give us a
twofer. They may well become standard-issue equipment for prospectors
and scientists alike. So navigation is an almost solved problem too. We are
good to go.

Once our prospectors tell us that there's enough of our favorite resource
to get us interested, we can start to ask, "Is the ore easy to extract?" This
is how we determine the fraction $f_{engineering}$ that I blithely set to 1 earlier on.
To answer the question there are a whole lot of questions we need to
answer that a terrestrial miner never has to face. Simply getting our mining
tools onto the asteroid is challenging. How fast is our asteroid spinning?
Or is it tumbling? We want asteroids that are simply rotating because they
are easier to land on; you put your spacecraft at one of the poles and start
it spinning at the same rate as the asteroid. From the spacecraft's point of
view the asteroid now appears stationary, so it can descend to the surface
safely. But if the asteroid is tumbling, then there is no stable pole. The
spacecraft will have to follow its chosen landing spot as it moves irregularly
below, blowing off precious propellant all the time. Remote observations
measuring the light curve of an asteroid can give us a good first idea. Following up with proximity prospecting will tell us far better how it spins.
    Even a simply rotating asteroid probably won't be that easy to anchor
to. There's a fairly unfortunate history of astronauts trying to dock with,
or at least grab hold of, satellites or rocket upper stages that have gone
out of control. NASA calls these "uncooperative targets."[35] In every case
the carefully rehearsed plan A did not work. Always the astronauts had to
fall back on improvising. When you're inside an aluminum can where any
hole can rob you of air to breathe, improvising is not the preferred course
of action. An out-of-control satellite could easily rip open your thinly

protected cabin. That these docking attempts eventually succeeded is a tribute to the astronauts' creativity under stress. Attempting the same thing when your uncooperative body is a large, irregularly shaped, lumpy, spinning, and even tumbling rock is more than a bit daunting. Moreover, because of the radiation hazards to humans in space, the mining in the asteroid's native orbit will have to be fully automated.[36] Autonomous distant robotic spacecraft are not likely to be as flexible and innovative as humans for a long time to come.

To deal with these new conditions we're going to need rugged mining spacecraft. That's an issue, as *rugged* is not an adjective usually associated with spacecraft. Mining in space presents unique challenges to two very different types of engineers. Asteroid mining will bring the delicate, minimalist designs of space engineers up against the tough approach of mining engineers. Reconciling their competing needs will not be easy. Culture clashes can be expected.

The Space Shuttle docked with the ISS at just 117 meters per hour, half the speed of a turtle, to avoid damaging either one.[37] This was sensible design, allowing both the docking spacecraft and the ISS to be as lightweight as possible to appease the Rocket Equation. Space engineers are taught to design their craft to be just strong enough, but no more, so as not to waste mass. That's also why they go for using as little power as possible. More power means larger solar panels to collect it, and that's more mass. Every tolerance is worked out carefully, with smaller margins than most engineers would be comfortable with. Designs are reworked numerous times looking for ways to save a kilogram, or even less.

In the unpredictable and far-from-benign environment of an asteroid, the highly optimized approach of space engineers won't work. Mining engineers have to build super-tough equipment for the unexpected—sudden hard lumps or gaps in the rock, for instance. That's not gentle on equipment. Any machinery that tries to mine an asteroid will have to be as rugged as mining machinery is on Earth. We know that most asteroids are lumpy, with many unpredictable voids within them. The machine making its way into the asteroid has to be ready to cope with sudden changes in resistance in all directions as voids are opened up. Seaplanes have a similar dilemma. They need to be both rugged and light. The pontoons they have instead of wheels have to be strong to survive in the water, bouncing on the waves at takeoff;

but they also have to be lightweight so the plane can get airborne and carry a useful cargo. This is one reason that seaplanes are a niche market, despite 40 percent of the world's population living near a coastline.[38]

Then there are new engineering challenges that just haven't come up before. For example, dust. Extracting the valuable ore from asteroid rock—called beneficiation—has to involve breaking up the rock into manageable chunks, and maybe grinding it into gravel or sand. That will generate a lot of dust. In microgravity that dust will largely hang around, settling only very slowly. If asteroid dust is anything like lunar dust, then it can be destructive. To quote a 2011 NASA report, "Lunar soil is highly abrasive and effective as a sandblasting medium."[39] It will be important to keep that dust out of the machinery.

On Earth we use water and air to sluice or blow the dust away, and gravity to keep it down. None of this works on an asteroid. New methods will have to be invented to control the dust. Doing that in the weightlessness, vacuum, and cold of space won't be easy. The mining site probably has to be enclosed. If the asteroid is not too big, putting the whole thing in a sealed bag may be the way to go. This will all have to be practiced, first on a small scale in a space station, then at scale on a captured asteroid.

Any dust that does escape the asteroid will gradually spread around its orbit. Since the first asteroids to be mined will be near-Earth asteroids, with orbits that cross that of the Earth, this debris will produce beautiful new meteor showers for us to admire. Any larger, pebble-sized debris, though, will create a new hazard for Earth-orbiting satellites. Asteroid miners may have to get liability insurance against knocking out a really expensive communications satellite.

Where should we do this beneficiation? It would be nice to bring the raw rock back to some large processing plant in Earth orbit. Then we could have big processing plants and people to oversee the machinery and fix it when it breaks. In Earth orbit a good supply chain can be established so that our expensive capital equipment can be used efficiently. This approach also saves on moving expensive automated mining equipment to the asteroid orbit and, quite likely, abandoning it there. Another suggestion is to crash the asteroid into the Moon and do the extraction there. This approach, though, would throw away the advantage asteroids have of having only weak gravity, wasting energy.

On the other hand, the Rocket Equation makes moving mass around the Solar System really hard. So much of the fuel is spent in just moving the fuel around, not the ore. Even an asteroid with an abundant resource like water is 90 percent nonvaluable rock. For precious metals 99.999 percent of the asteroid is nonvaluable, at least in this first round of mining. That leads to one simple conclusion: most of the beneficiation will have to be done in the asteroid's native orbit so that far less fuel is needed to bring it to market. The Rocket Equation also pushes us to make the mass of the robotic mining equipment a lot less than the mass of the ore being brought back, and very much less than the mass of the asteroid. If we could refuel our rocket for the return journey out at the asteroid, we could bring back a lot more ore.

If you could stop the asteroid's rotation, that would make it easier to work with. But to do that you need to get a firm grip on it. That's not easy to do for a rubble pile, which is what most small asteroids seem to be. What you grab onto is just a boulder, not the whole asteroid. If it comes loose, as seems likely, then you really haven't de-spun the asteroid at all. One suggestion, as we saw, is to capture the entire asteroid in a bag. If you can do that and cinch the bag tight around the asteroid, compressing it on all sides at once, then you'll have a strong enough hold on the entire body. Then you can fire some rockets to stop the spinning. The NASA Asteroid Redirect Mission study calculated that it would, to my surprise, take only a small amount of rocket propellant to make this happen. So it's not a crazy idea. The bag has to be hard to rip, as the surface will be irregular and may be covered with sharp-edged rocks. Fortunately, thanks to the asteroid's low gravity, it doesn't have to take anywhere near as much force as it would to lift a house with a net here on Earth.

The next step will be digging into the asteroid to mine rocks for your refining equipment. Gravity is really weak on all asteroids, because their mass is so small compared with Earth's. Typically, their gravity is a million times less than Earth's, so we call it microgravity. (Just as a micrometer is a millionth of a meter.) This microgravity is a major reason that we want to mine the asteroids in the first place, rather than the Moon or Mars. It makes fewer demands on our rocketry if getting off the surface needs something more like a strong jump rather than a big rocket. But microgravity is not always helpful. How do you hold on firmly enough to dig into the asteroid? The problem is conservation of momentum once again. Digging means we have to push down hard into the rock, which means our digging tool gets pushed back. In this low gravity it could well float off into space.

Worse, the low densities of asteroids show that they are not solid rock. Your ore-bearing asteroid could well be a rubble pile. You can't then just attach to a rock and assume that you have grabbed onto the entire asteroid. It will likely come away in your (robot) hand. And if you grab another and another and another, they will all come away too. In fact, you can keep going until there's nothing left. Maybe that is okay; you could grab one boulder at a time and put it into your ore-extraction machine, leaving the asteroid spinning underneath you. It would be much simpler, though, if you could attach your mining machinery to the asteroids.

Rubble-pile asteroids came about because of the rough and tumble background of asteroids early in the Solar System's history.[40] Collisions broke up not only the planetesimals but also the fragments from those first collisions. Most of the fragments collided many times over. If the crash was fast enough to break up the fragment into smaller fragments, but not quite fast enough to blow the whole pile apart, then the smaller fragments would slowly rain back down together and form a pile of rocks. Rubble-pile asteroids are held together only by their very weak gravity and some similarly weak chemical forces. Rubble piles qualify as one type of granular material.[41]

Granular materials include sand piles, grain in a grain elevator, or cereal in a dispenser at a breakfast buffet. How granular materials behave is the topic of a fairly new area of research called granular physics. I confess I had never heard of granular physics until 2011. (We astronomers live quite cloistered lives.) But then I was lucky enough to be at a workshop at the Aspen Center for Physics. This is a unique place in several ways. For one thing, Aspen is a very beautiful place set in the verdant Roaring Fork Valley and surrounded by the peaks of the Rocky Mountains. It also has great restaurants for the rich clientele who flock there every winter. Physicists can afford to go there only because we get special off-season rates; we are also pretty popular with landlords because we rarely cause damage to their quite spiffy apartments. Occasionally being boring comes in handy.

One of the ways the Aspen Center for Physics mixes things up is to make you share an office with a person who researches something utterly different than you do. That's how I met Karen Daniels. Karen was sharing an office with an astrophysicist friend of mine, Andrew King. One day I was talking with Andrew in his office about my newfound enthusiasm for asteroid mining. As I explained to him how most asteroids were rubble piles, his roommate, whom I admit I hadn't really noticed till then, started perking up and paying attention. Eventually she broke in and asked questions, and then more and more of them. Soon she started explaining granular physics to us, who knew nothing about it. It turns out she is a leading researcher in granular physics, but she hadn't heard that most asteroids were her kind of rocks. We both learned a lot over the course of the next few days.

Most of the ideas for anchoring to an asteroid involve harpooning it in some way. Karen explained that this probably isn't a good idea. Rubble resists penetration by a harpoon far more than you'd guess. That's because rubble piles tend to "lock up." That is, they tend to form arches that block the movement of the rocks. Lock-up jams form when there's a strong force pushing down on the rubble. This is how the cereal can get trapped in those breakfast dispensers (figure 8). It is also why "bunker-busting bombs" are less effective at destroying underground bunkers than you'd guess. The force of their explosion compresses the rock below, making it lock up, so protecting the next layer of rock from being blown out. This could, and probably will, happen in space too. Anyone who wants to push on an asteroid to deflect it from hitting Earth has to push it with a pretty strong force. You'll want to know what the result will be when you do that. On a rubble pile that effect may be very different from what you expected.

The sharper your harpoon, the less rock it has to displace when it pushes through the surface, and so the easier it is to penetrate the surface. With that in mind, a group in Japan led by Takeo Watanabe of Teikyo University is taking a lesson from samurai swords.[42] There are stories of how these swords can cut through metal and stone, which sounds like just the thing for asteroids. Their secret lies in the special *tama-hagane* steel developed a thousand years ago by Japanese sword makers. Normal, stainless steel blades deform on very hard surfaces. The Japanese researchers hope that tama-hagane steel will not bend. There have not been many scientific studies of this steel. Presumably the owners of historic samurai swords are loath to subject them to any test that might break them, and obtaining

Figure 8: An example of lock-up, a special property
of granular materials, is illustrated by this breakfast
cereal dispenser. Asteroids made of rubble may
behave the same way when hit.

new tama-hagane steel is very expensive. Watanabe has a small sample and
hopes to test it soon. One of the possible uses for quasicrystals is to make
extremely strong blades. Maybe they will give samurai sword steel some
competition!

Karen came up with a completely different approach to anchoring to a
rubble-pile asteroid, far subtler than trying to harpoon one. She wants to
insinuate a whole system of "roots" into the rubble. They will behave just
like the roots of weeds in your garden that make it hard for you to pull them
out, even though the soil the weeds grow in is a pile of quite disconnected
particles of dirt. This sounds like just the thing for anchoring to asteroids.

There's only so much Karen can do to test this idea on Earth. There are
drop towers that give a few seconds of milli-gravity, including one 110

meters tall at the University of Bremen that gives just 4.5 seconds of free fall. An airplane on a parabolic flight (please don't say "vomit comet," the operators beg) gives several 30-second bursts of micro-g, while a suborbital rocket, such as Blue Origin's *New Shepard,* gives 10 times longer, about 5 minutes. Karen has already flown on a series of zero gravity parabolic airplane flights to test out her ideas. Any experiment that takes more time will have to be done in orbit. Insinuating roots into rubble, for example, is inherently slow. Using the ISS for the first small-scale tests seems like a natural.

Another strange feature of granular materials may become really important. The speed of sound becomes very small in them. Only a tiny knock is then enough to produce shock waves. As just a few granules are taking the strain, a shock could destabilize the whole asteroid, causing it to rapidly alter its shape. If the asteroid is rotating this change in shape could cause it to wobble badly, which is unlikely to be good for your mining equipment. In a worst case the asteroid may act like the row of steel balls in a Newton's Cradle; a knock from the first one doesn't move the middle ones, but the last one flies off. That's not the effect we were looking for.

The advent of asteroid mining, and of planetary defense, means that these fairly abstruse research topics about how granular materials behave in microgravity are about to become practical issues. "More research is needed," we scientists always say. Here that is not only true, it's urgent, because these studies take years.

Once we have our mining machines in place, there's the actual extraction of the resource, the beneficiation. How do we refine the water or precious metals out of the raw material in the difficult environment of deep space? There's not yet much work being done on this really basic question. We need to find ways to do this that don't depend on having gravity, can work in a hard vacuum, and are happy with extreme cold. John Lewis, who wrote the pioneering book *Mining the Sky* to advocate for space mining, has listed several methods that should work in space.[43] None of them have been tried out in zero gravity though, let alone at an asteroid. One big limitation is power. On Earth most mining starts with crushing the rock into small pieces. That takes a lot of power, while space missions have had to make a point of getting by on as little power as they can. The *Chandra*

X-ray Observatory, one of NASA's Great Observatories, runs on about as much power as a hair dryer.[44] This constraint comes about because solar panels are too heavy to launch a much bigger area of them. So that's a challenge. Joel Sercel's company TransAstra, based in the Los Angeles area, is working on using sunlight to directly heat water-bearing rocks to "boil off" the water and then collect it. Instead of photovoltaics, the company uses solar furnaces and clever new "non-imaging" mirrors to reach the very high temperatures needed. TransAstra calls its version "optical mining™."

There are many techniques to separate out high-density ore like iron and precious metals. They have some great names: shaker tables, spiral concentrators, and dense water cyclones.[45] Most methods are based on using gravity, but luckily not all of them. Instead some use centrifugal force. Panning for gold works this way. They should be modifiable to work in zero gravity. Other heat- or chemistry-based technologies may be better at extracting the ore than simple crushing. Platinum beneficiation uses some pretty heavy equipment on Earth. The nickel-carbonyl processing for platinum may help. It is another method we can start testing on the ground.[46] This method picks up all the nickel and iron and leaves behind the cobalt, platinum group metals, and gold. It is also a circular process— at the end you are left with the same chemicals you started with. This is a big plus for space mining as it cuts down how much mass has to be moved around the Solar System. Like sending roots into rubble piles, these methods require longer times than suborbital flights can provide. So orbital tests would be essential. Most laboratory-scale and bench-based metallurgical testing uses kilogram-size samples, which is a good size to use at the ISS.

There are some even more creative ideas out there. One is biomining the asteroids. In 2016 a group of Japanese scientists discovered some bacteria that had evolved naturally to eat plastics.[47] It turns out that bacteria can be bred to eat a remarkable variety of substances. Biomining on Earth is a new way to attack the problem of waste recycling, especially for difficult items like obsolete electronics. Jesica Urbina, then at the University of California, Santa Cruz, studied exactly this with her PhD advisor, Lynn Rothschild of NASA Ames. Rothschild studies bacteria that can survive in extreme conditions, "extremophiles," because she is interested in exobiology. She wants to find the highest temperature life can deal with as a guide to whether an exoplanet can host life. Biomining the asteroids is a possible spin-off of her work. Even before the discovery of plastic-eating bugs, a group led by Michael Klas at the University of New South Wales in Sydney had proposed biomining carbonaceous asteroids to make methane. The researchers suggest using archaea bacteria, which naturally produce methane. That gas can then be used as rocket fuel. The appealing thing about biomining to an asteroid miner is that it's a way to beat the Rocket Equation. The bugs reproduce themselves exponentially, growing off the material of the asteroid. So your spacecraft needs to carry only a small feedstock to "inoculate" the asteroid, as Michael Klas calls it, not a huge mining machine.[48] Maybe "infecting" would be more accurate? A solid nickel iron asteroid may not have enough nutrients for bacteria, while normal silicate rocks can suffice for some bacteria. There is iron in stony asteroids, just much less than in the metallic ones. Biomining might make them a good place to begin asteroid mining after all. That would make finding ore-bearing asteroids a lot easier. Lynn Rothschild is part of a team looking at which bacteria could be suitable and could give a good return on investment on the Moon or Mars.[49] So far one of the four microorganisms they investigated, *Shewanella oneidensis*, looks promising.

Biomining can get a start by testing different bacteria on simulated carbonaceous asteroid material in Earth-based laboratories. The tiny amount of carbonaceous meteorites in our collections means that simulated meteorites are all we can practice on. The curators of real ones naturally consider them far too valuable to use up on experiments that seem so far-fetched. Their understandable reluctance means that it can take months to get even a gram-sized sample. Biomining experiments will need a lot more. Life can exist in extraordinarily tough conditions. An ever-wider array of extremophile bacteria is being found. Still, the asteroid environment goes beyond even what these organisms can handle. It may be possible to force the bacteria to evolve toward these tougher conditions. Or, Klas and his colleagues suggest, we may have to relent and provide some kind of tank on the surface of the asteroid with milder conditions where the bacteria can do their work.

On second thought, a bioreactor tank is a safer route. If the bacteria can eat the wide range of organic compounds that make up a carbonaceous asteroid, they might be able to eat a lot more. Space suits, for example, or people. Unless you are writing a space horror movie script, you really want to keep that sort of bug away from everything that it isn't supposed to eat. We will be keeping them in a safe environment, not letting them roam freely over the asteroid eating it all up.

After extracting the valuable 10 percent of water, or 0.001 percent of precious metals, we'll be left with a huge amount of less valuable material. What will we do with the bulky mining tailings? There is an economic incentive to keep them all together. Thousands, even millions, of tons of rock or iron in space is a valuable resource in itself. It is far more than we could launch from Earth with rockets at any reasonable cost. We just have to get them somewhere that someone can use them, so we can sell them. Until we can do that it would make sense to store our new property somewhere. It's a bit like "land banking"—owning a property, but not doing any development on it until it becomes worthwhile. There are a few options for how to do this. We could separate the tailings into mineral types and store them in the asteroid's native orbit for later use in loose form. We could use volatiles to embed them in a matrix, making a "faux asteroid." That may need water, though, which would eat into our primary profit stream. We could sinter the ore into blocks using a solar furnace. Or we could just keep the loose tailings in a big bag. Who knows? They may come in useful someday.

It's no use extracting a pile of precious ore if we can't get it to somewhere that we can sell it. Right now that means back on or near to the Earth. How do we bring it all back home? There are two parts to that: propelling our ore back to near-Earth space and getting it back onto the Earth without burning up in the atmosphere. We need to bring back enough ore to make the hefty profit needed to justify the whole enterprise in the first place. We saw that with 20 metric tons of ore at $50 million a ton we'll be realizing $1 billion in sales. Twenty tons is 2,000 times more mass than any mission has tried to return from an asteroid so far.

If we had better rockets, we could transport larger masses around the Solar System. There are several ideas in the works to give us those better rockets. What's happening with rocket science?

One way to get rockets with more oomph is to keep scaling up the SEP. The SEP engines developed for the Asteroid Redirect Mission are now available to purchase from the Aerojet Rocketdyne Company in California.[50] Scaling them up in power, maybe simply by having banks of them, would be needed if we are to start bringing back tens or hundreds of tons of ore.

Another approach is to refuel our rocket with something that the asteroid can provide. For the mining of precious metals this scheme is not going to work so well. They are concentrated in asteroids with very little water or organic materials. Instead this method should work well for any carbonaceous asteroids that we have mined for water or methane. Either one could be used to refuel our rocket. We only have to be sure we don't need so much that we rob our profits. Turning water into fuel is pretty complex. The water needs to be purified, separated into hydrogen and oxygen, and then both gases must be liquified. Only then can you refill the tanks of the rocket. It is something we can't yet do in Earth orbit, let alone at a remote asteroid. Water in its liquid form is far easier to handle. Momentus is a company founded to make a new type of water-based rocket.[51] Joel Sercel, whom we just heard about in the context of "optical mining™," is also chief technical advisor to Momentus. He calls the technology "electrodeless plasma propulsion," and it is a design he came up with for his Caltech PhD in the early 1990s. Early sales are promising.

We could revisit nuclear power for propelling spacecraft. We know fission reactors work, and they provide lots of power. Plenty of people are nervous about using nuclear technology in space. On the other hand, space is already full of dangerous radiation. A bunch of nuclear rockets won't add noticeably to that radiation. While nuclear weapons are banned in space under the Outer Space Treaty of 1967, nuclear reactors are not banned. The Soviet Union used to launch them routinely, although they fell out of favor after 1978 when *Kosmos 954,* which was carrying one of these reactors, crashed in Canada. Canada sent the USSR a $15 million invoice for the cleanup; the Soviets paid up, but only about half the bill.[52]

Lightweight reactors for space use could reasonably be ready in a decade or so. Fission reactors break up heavy atoms, such as uranium 235, releasing large amounts of energy in the process. After many years of being quiet on the topic, in 2015 NASA restarted work on what it calls a "kilopower" program to build a fission reactor that could generate 10 kilowatts or more of power for a decade or more, to use in space. The agency plans to use a reactor design from Los Alamos National Laboratory's David Poston. His design is unusually compact and is engineered so that a meltdown can't happen. The test reactor, which passed its tests in 2018, had the great name KRUSTY—kilopower reactor using Stirling technology—and cost only about $20 million, which is small change for NASA missions. Poston says the KRUSTY design could scale up to a megawatt.[53] That's 10 times what the advanced SEP can give us, and it doesn't need huge solar panels either. It will still need large cooling fins to dissipate excess heat from the reactor. NASA likes the idea of using these reactors to power bases on the Moon or on Mars. They'd also be great for powering missions beyond Mars, to the outer Solar System, beginning with the Main Belt asteroids. NASA and Los Alamos are now hoping to move on to build a reactor for an actual spaceflight.

More out there is the idea of fusion rockets. Fusion reactors combine light elements into heavier ones. These reactions release even more power than fission reactions. The Sun is powered by fusion reactions at its core, converting hydrogen into helium. But despite decades of attempts to build fusion reactors on Earth, no one has yet succeeded. That lack of progress got some scientists at the Princeton Plasma Physics Laboratory thinking about what they might do that could get some quick results. They found that a reactor that was open at one end is much easier to build, if you choose the right reaction. A reactor open at one end is a rocket. They call their design direct fusion drive. They have teamed up with the company Princeton Satellite Systems, which will design everything else surrounding the reactor that is needed to make a fully operational fusion rocket.[54] The Princeton Plasma Physics Laboratory will develop the reactor. The researchers claim that for $70 million they could have a working prototype within a decade. If they do, then their engine could revolutionize Solar System transport, including returning large masses from asteroids.

In the movies, betting on a breakthrough, such as fusion reactors, always works—and quickly. The history of fusion, however, suggests that we shouldn't pin our hopes only to this idea. As the advertised price tag is less than one of NASA's "small explorer" missions, that would seem like a good side bet.[55] In fact, NASA has given some grant funding to the team. We'll see what happens.

Lastly there is orbital mechanics. This is the study of how to get "there and back again" in space. Orbital mechanics is about designing trajectories to move our 20 metric tons around the Solar System from the native orbit of the asteroid to an orbit around the Earth, or down to the Earth's surface. Right now, we can only get a 2-ton spacecraft out to the asteroids that are the easiest to access. Coming back is not "downhill." Getting down to LEO or to the Earth's surface needs the same delta-v as leaving. We don't want to crash-land, so we have to slow down. We don't necessarily have to come back all the way down to Earth, though, to the bottom of its "gravity well." Twenty tons seems challenging, but not off scale. The Asteroid Redirect Mission would have brought back about 5 tons to a high orbit, which needs much less energy. Asteroid water has double the value in a high Earth orbit than in a low orbit, so that's where we would prefer to sell it, as long as someone wants to buy it there. But the first markets are likely to be down in LEO. How could we get a big mass down there? There are some good ideas.

A straightforward way to get more energy is to take it from somewhere else. The Moon's gravity and the Earth's atmosphere are two well-established methods. A fly-by can use the Moon's gravity to slow down our returning ingot of ore. This can gain us about 1 kilometer per second, 2 if we swing around the Earth and go by the Moon again.[56] Compared with the 12 kilometers per second or so that our ore will be returning at, it isn't much, but it is free and ought not to be ignored. We can also slow down our ore shipment by skimming through the Earth's atmosphere. This is called aerobraking. It's tricky. Aerobraking means coming within 100 kilometers of the Earth's surface at high speed. That may give governments pause before they grant permission for such an attempt by a mining company. A slight mistake could turn aerobraking into "lithobraking"—being stopped by rock at zero altitude—all too easily. Using the Moon's gravity to slow down instead won't cause the same concerns. There's no one to hit on the Moon, as yet.

One idea for a very energetically easy return trip comes from Marco Tantardini, an Italian entrepreneur who sparked the study of ARM. He suggests using some special low-energy orbits called invariant manifolds to return from the asteroid. This could let us bring back a lot more ore. But these orbits take longer to get your payload back home. The ARM spacecraft would have taken two years to return a 5-ton boulder to a lunar orbit, using a 40-kilowatt solar electric engine. Five tons of ore might be worth $250 million, so at the low end of profitable.

The best trajectory to use to bring your ore back is not just a question of orbital mechanics. Economics matters too. Time is money, after all, so a quick route that is energetically costly (higher delta-v) may be a better choice than a slow but low delta-v path. The companies will have some tricky calculations to make: some money now, or lots more later? We certainly have to bring back a lot more mass than we sent out in our mining spacecraft if we are to make a suitably huge profit. Maybe the first, highest-value ore will come back by the fast route to pay off the entrepreneurs' bank loans, and the bigger follow-on ore shipments will take their time but will eventually make the investors very rich.

We don't always want to bring our ore to the ground, but when we do, we will have to find a way of coming through the atmosphere safely. Coming through the Earth's atmosphere is usually called reentry, which was correct for spacecraft launched from the Earth. The asteroid ore never left Earth, though, so technically it is just "entry"; I doubt this will catch on. Getting the ore to the Earth's surface from space poses some problems. The ore will be coming back toward Earth at high speed, about 12 kilometers per second, or over 43,000 kilometers per hour. To put that in perspective, 40,000 kilometers is the circumference of the Earth. As our ore speeds through the atmosphere it will heat up to red-hot temperatures as it slows down. We really don't want our precious ore to boil away in the last brief leg of its long trip, imitating (spoiler alert again) the movie *The Treasure of Sierra Madre*. A neat answer would be to bring along some nonvaluable rock to use as a shield against the heat. This rock would burn away during that short trip through the atmosphere, so it's called an ablative shield. If we use up too much of our payload on this shield it will eat up our profits. An alternative is to slow down high up in the atmosphere where the air is less dense, so the spacecraft stays cool. The rarified atmosphere means that the spacecraft needs to present a larger area to the air to slow down. This could be done by inflating a balloon or by trailing a parachute-like "balute" behind it, as proposed by Radu Dan Rugescu of the Polytechnic University of Bucharest and others.[57]

We'll also need to keep careful control over our spacecraft as it plummets through the atmosphere lest we lose it in the ocean or cause damage to those below. That will probably demand that the ore ingots from the asteroid not be irregular in shape. If they were, then they would tend to tumble when the atmospheric pressure builds up in front of them, making the landing spot uncertain, as well as exposing the ore to extreme heat. The miners will have to pack their paydirt carefully for the return trip. Perhaps zero gravity can be used to naturally turn the ore into a spherical blob if it starts off liquid.

Controlled entry of large masses has, in fact, been done many times. The Space Shuttle weighed about 100 tons and was brought back down safely over 100 times. There was just one disastrous failure when, in February 2003, the space shuttle *Columbia* disintegrated while reentering because its heat shield developed a gap. The crew of seven was lost. Even then the rain of debris over Texas didn't hurt anyone else and caused only minor property damage.[58] So perhaps bringing tons of precious metals hurtling through the atmosphere is a less scary prospect than it sounds. True, that payload will be going faster than the Space Shuttle did, but if we keep to ore chunks of 10 tons or so they will have less energy than the shuttle did. That may be perfectly acceptable, and we can land in one of the many areas less densely populated than Texas to reduce the risk to people further still. Nonetheless, the mining companies will have to show that they can reliably target those areas and have a backup plan in case something goes wrong.

We have seen, then, that the *means* needed to turn asteroid mining into a real and profitable industry are available. In some cases, there are lots of ideas competing to remove the barriers; in other cases, there may be only one. But none of the barriers we can see now appear insurmountable. With one crucial exception: will we actually be able to make a profit selling resources mined from asteroids? Right now, no one knows. But new developments are converging to turn this once fanciful idea into an emerging opportunity. *Opportunity* is what we look at next.

PART

IV

Opportunity

8

Getting Space off the Ground

Now we know what the asteroids could mean for our three motives: curiosity about the universe, what we have to fear from them, and how they might make us rich. Just how are we really going to fulfill any of these longings? Now is a good time to begin working toward all these goals. Space is wide open. There are a whole slew of changes happening in the space industry that could provide the opportunity for us to start mining the asteroids. They range from new technologies through new ways of doing business to new industries in orbit. Taken together, this NewSpace lets us sketch out a plausible path from here to a full-fledged profitable industry that makes use of space resources. And we need to make profits from space if we are to reach our goals.

Space already has a sizable economic impact. Worldwide in 2018, space was a $360 billion a year industry, according to the annual *Global Space Economy Report* from Bryce Space and Technology, whose CEO is Carissa Christensen.[1] That sounds impressive. Yet it is only about 1/2 percent of the world economy and is less than the revenue of the giant retailer Walmart. Around a third of the total space industry comes from "public goods," those provided by governments but now utilized for profit by businesses. Before continuous high-quality weather satellites, forecasts were pretty hit or miss. Now they can be trusted several days in advance. This is due to a combination of vastly better computing capabilities and the vastly better data that weather satellites give us. Weather data is supplied free of charge and already processed into predictions by governments.

This system works. It saves lives with accurate hurricane tracking, among other things, and gives a great boost to the economy by letting us plan around bad weather. The Global Positioning System (GPS) is also a public good. Originally a military tool, GPS is now ubiquitous, essential, and free. GPS-based satellite navigation has been coupled with smartphones and has given us Uber, Waze, and dozens of other apps. Its accurate clocks are used by banks to time stamp transactions. GPS and its siblings are so big that a quarter of the space economy lies just in GPS chip making. That's not what we normally think of as a space activity, yet really it is one. The rest of space activities are truly profit-making commerce. The biggest part is made up of communications satellites, mainly providing broadcast TV, which accounts for nearly another quarter of the space economy. The business most obviously about space is making rockets and satellites, but that accounts for only about a tenth of the total. Our problem is that nothing commercial is yet built in space. There is no industry actually in space. That limits the customer base for asteroid-supplied materials.

What are the prospects for commercial growth in space? There are new space-based industries now emerging, from small satellites for imaging the Earth daily, or even hourly, to global high-speed internet access. Space tourism, research, and even manufacturing are on the horizon. This radical expansion of our space activities may create a demand for space-derived resources. How is any of this going to get off the ground?

Launch is the sine qua non of space—if you can't make it to orbit, nothing else matters. The big problem with launch has been the high cost of a ticket to ride to orbit. The reason for the high price is that all the rockets ("launchers" in space lingo) built until recently were "use once and discard." When you are talking about tissues or staples, that's really a fine way to go. But when your product is several hundred million dollars' worth of high tech, it's really pretty dumb. The usual comparison is to imagine that you and 300 friends take a $300-million Boeing 787 from New York to Singapore and abandon it there. The $1 million price of your ticket for that one-way flight would make today's first class seem like a bargain basement price at about $15,000. (Going via private long-range jet may cost a similar amount.)[2] Space is not inherently so expensive. The fuel cost for one of SpaceX's *Falcon 9* rockets is just $200,000, according to Elon Musk, its founder.[3] That's under 1/2 percent of the total. Launch costs could be 50 times cheaper before fuel costs reach the same fraction of your ticket cost as they do for airlines today.

Just how expensive have rides to orbit been? The 1960s Apollo *Saturn V* rockets put tonnage into orbit at a cost, in today's dollars, of around $10,000 per kilogram.[4] For an average weight person, that's a $2 million ticket. Actually, it's more like $4 million, if you take along enough equipment to keep you alive, which is advisable. After 50 years of the space program, the cost of launching to orbit in 2010 was well over $10,000 per kilogram, using the Space Shuttle.[5] This enormous cost has kept the space industry small. It's hard to find businesses that can make a profit with transport costs that high. That explains why the year 2001 was nothing like the movie *2001: A Space Odyssey,* with Hilton hotels in giant rotating space stations and regular passenger flights to orbit on PanAm (a now-defunct airline). No one could have afforded it. Luckily, that is all changing.

We can be optimistic about launch costs dropping, because they already are. SpaceX has cut the cost to get cargo into space by a factor of seven.[6] True, it had some government funding at crucial stages. SpaceX is not alone. Amazon founder Jeff Bezos's rocket company Blue Origin is doing the same. And the "old but rejuvenated" aerospace regulars, United Launch Alliance and Arianespace, have realized that they too have to embrace reusable rockets or they will be driven out of business. It was surely no coincidence that they both announced their reusability plans just months before the first landing of the first stage of a SpaceX *Falcon 9* rocket. Competition concentrates the mind wonderfully.

Let's think about ticket prices to orbit. Flying in a Russian *Soyuz* spacecraft has gradually cost NASA more and more, from about $30 million in 2010 to just over $90 million in 2019.[7] That's not cheap, but a few private individuals have ponied up a similar price at the earlier discount rates. Even the early flights of *Dragon* and *Starliner* from SpaceX and Boeing will cost similar amounts. Not everyone finds that too expensive. Blockbuster action movies can cost $300 million, and their stars can make north of $10 million.[8] We shouldn't then have been surprised that movie star Tom Cruise was the first to announce a *Dragon* flight to the ISS as part of making a movie.

For most other potential customers, that's still pretty expensive. But the cost can come down, especially once the development costs are paid off. SpaceX has a "private crew program" and has partnered with Peter Diamandis's Space Adventures to promote it.[9] But they don't post ticket prices. Not to be outdone, Boeing has an agreement to fly crew to Axiom's addition to the ISS in its *Starliner*.[10] The Crew Dragon capsule that flies on top of the *Falcon 9* is designed to take seven people to orbit. One of them would surely be crew, so the launch cost is about $10 million per person. That's not the ticket cost, as there's also the cost of the *Dragon* spacecraft, plus all the launch facilities to pay for, and a profit for the operators, but it suggests that quite a price break is possible.

As most of the rocket and the *Dragon* are reused, the incremental cost of a new flight is just the fuel, the refurbishment cost, and the unretrieved upper stage, plus the cost of the launch crew. Reusing the rocket and spacecraft 10 times, as SpaceX plans to do, could lower the ticket cost to the low millions. That's hardly cheap, but it's only a few times what hundreds of people have already paid up front just to experience five minutes of weightlessness with Virgin Galactic and Blue Origin.[11] It is a plausible price, too, for small countries that want a space program but had been excluded by the high cost barrier to entry. Large corporations too have research budgets that could handle million-dollar tickets, if they can point to potential benefits down the road. It seems to be a price point that opens up space to many more customers and can plausibly be matched by the other providers. (By the way, when you buy your ticket to space, it would be advisable to read the small print. You want to be very sure that the price includes sandwiches—and oxygen! Not to mention a ride home.)

The eye-catching, breathtaking part of the SpaceX rocket system is the controlled vertical landing of the big first stage of the *Falcon* rockets. Reusability was not required by their contract with NASA but was self-funded by SpaceX as a strategic move. The big first stage of the *Falcon 9* returns either to the launch site or down range onto an autonomous oceangoing barge. The Atlantic barge is called *Of Course I Still Love You*. This odd name is in homage to the science fiction book *The Player of Games* by Iain M. Banks. Its Pacific Ocean twin is called *Just Read the Instructions*, from the same book. The first successful landing of a *Falcon 9* first stage took place at Cape Canaveral on 16 December 2015. The first seaborne landing success followed just a few months later on 8 April 2016. Now there have been dozens of first-stage landings by *Falcon 9* . The spectacular test launch of the larger *Falcon Heavy* rocket on 6 February 2018 featured the simultaneous landing of two first-stage boosters side by side at Cape Canaveral.[12] At the time it looked like science fiction. But as Chris Lewicki, former CEO of Planetary Resources, says, "Everything is science fiction right up to the point that it's science fact."[13] Reusability has become routine.

To be fair, Blue Origin had successfully landed its entirely self-funded *New Shepard* rocket a month before SpaceX, in November 2015.[14] That was a dramatic and impressive achievement. The Blue Origin rocket was not meant to reach orbit, only to enter space briefly on a parabolic "hop." That means that the energy and velocities involved were much more moderate. At the time it looked like Blue Origin had been wise to choose the easier problem to solve first. After all, SpaceX had experienced nine failed landing attempts over three years. When SpaceX had its success, though, it marked a major change in the economics of launching satellites.

Concentrating on just retrieving the first stage makes sense. SpaceX says that 80 percent of the cost of a *Falcon 9* lies in the first stage. This is mainly because 9 out of the 10 Merlin engines on a *Falcon 9* (and 27 of the 28 engines on the *Falcon Heavy*) are in those first stages. As the engines are the most complex and costly parts of the rocket, recovering 90 percent of them (or 96 percent for *Heavy*) can save a lot of money. This does mean that the engines and the rocket have to be engineered for reuse from the get-go. That's one reason they are not pushed to maximum performance. Better to keep them structurally sound so that they can be sent back up on another mission.

Reuse will also lead to reliability. Being able to inspect an airplane engine after a flight makes it a lot easier to find and fix any flaws it may have. That has to be true for rocket engines as much as for jet engines. That surely lies behind SpaceX's confidence that its "block 5" *Falcon 9* first-stage boosters can be reused 10 times. It had lots of used rockets from earlier blocks (designs) to inspect and learn from. As of this writing, SpaceX is up to seven reuses of a booster.[15] The servicing of *Hubble* is an example of how getting your hardware back helps to diagnose problems. It allowed the reason for the failure of *Hubble*'s gyroscopes to be discovered once the broken ones were brought back to the ground. No one else had ever had that ability to learn the weak points of rocket designs, except for fragments collected from RUDs. ("Rapid unscheduled disassembly" is the industry term for a rocket explosion. Yes, it's tongue in cheek. Even nerds have jokes.)

If you could reuse the rocket and spacecraft 100 times, as SpaceX has claimed is possible, then the simple-minded cost exercise above would come out at about a $100,000 ticket price. This may not be fanciful. SpaceX claims that the new *Starship/Super Heavy* system could fly 100 people to orbit (or to anywhere on Earth) for a cost to SpaceX of just $2 million. That's only $20,000 per passenger, comparable to long-haul, first-class plane ticket costs today.[16] SpaceX will no doubt add a hefty profit margin to that, but it would still be a game changer.

Going from 10 to 100 uses is a tough engineering challenge. Maybe, though, it is less of a challenge than making the first reusable rocket and spacecraft because it is no longer breaking a paradigm, just stretching it. When jet aircraft first flew from New York to LA in the early 1960s, it was a big deal to make the trip because a ticket cost five or more times the price of the same ticket today, compared with average incomes. And it was risky. Airplane crashes were common.[17] Instant insurance booths were a normal, and somewhat unnerving, sight at airports. My father started taking annual transatlantic trips in the 1960s and each time he left he would say good-bye as though it might be the last time we would see each other. Yet he went. (He was fine.) Passengers to space will make similar judgments. Those 1960s airplanes used fundamentally the same jet engine technology as we do today. It is incremental improvements to that technology that have cut prices to everyday affordability, driven by competition. Jet engines, after decades of enhancements, are now some of the most reliable machines in the world. Pushing rockets toward airplane standards and flying them 100 times or more no longer seems fanciful. Best of all, SpaceX has competitors. Blue Origin has its own partly reusable orbital rocket, *New Glenn,* under development, and ULA and Airbus/Arianespace both have new rockets about to fly. All of them stress low cost and customer service.[18] Competition always drives customer prices down. The *Falcon Super Heavy* and its *Starship* passenger-carrying upper stage will be fully reusable, which is a big part of why SpaceX believes the ticket price can come down a lot.

If we could get to space more cheaply, then our spacecraft, mining machinery, and space telescopes (my personal passion) could be built less like greyhounds and more like bulls; less like a delicate mechanical wristwatch, more like a truck. That way we could make them cheaper. If they were cheaper, they wouldn't have to be so perfect because we could just replace a failed satellite with a new one. (That would not apply so much for a crew carrier though!) Perfect is expensive. Abandoning that ideal would make for a virtuous cycle of downward-spiraling costs. Getting this cycle going has been the problem.

NASA had a big part in making cheaper launches happen, thanks to innovators inside the agency.[19] In 2005 NASA administrator Mike Griffin persuaded the George W. Bush administration to set aside $500 million a year to start the acquisition of transport services to the ISS from private companies.[20] NASA set up an office at the Johnson Space Flight Center in Houston. Waggishly, it was called C3PO—the Commercial Crew & Cargo Program Office, in a clear nod to the droid in *Star Wars*. In 2009 a new NASA administrator, Charles Bolden, was enthusiastic about keeping the program going, as was Lori Garver, the deputy NASA administrator. They pushed hard to bring this new commercial approach to spaceflight to NASA. This was not a popular initiative. Ashlee Vance, in his biography of Elon Musk, quotes Garver as saying that the opposition was so intense that she got "death threats and fake anthrax sent to me!"[21]

It was not, in fact, an obviously good idea. NASA had a system it was comfortable with and changing it could have been a huge mistake. True, the U.S. Congress had been urging a more commercial approach from NASA for some time. Still, that's a big gamble for anyone to take, especially a government official. The arguments had to be good. Partly the move was a response to the retirement of the Space Shuttle announced in 2003, which actually took place in 2011. This meant NASA would have no U.S.-based way to get either supplies or crew to the ISS. With the limited budget available, the old method could not produce a successor capability. Bill Gerstenmaier, then the associate administrator for human exploration and operations at NASA, explained that NASA had used a new approach due to these budget concerns. In a sense there was no good alternative. Nevertheless, it was a gutsy move.

The way this revolutionary approach worked sounds a bit dull. It involves contracts. Yet it was probably the most consequential government action on space since the decision to build the ISS. The idea was to use Space Act Agreements. These were a type of contract NASA had long been authorized to use but had not tried. Normally NASA had bought services under the Federal Acquisition Regulations (FARs), which allows the agency to tell the company precisely what to build, how, and when. A Space Act Agreement is much more flexible.[22] It is more like a normal purchase you would make in a store. NASA agrees to pay a certain amount for a particular service. The company then bears all the risk of failing to make a profit. The old FAR system instead had been "cost plus." These contracts guaranteed a profit over and above the costs, so NASA bore the risk if the company's development costs grew too big. In the early days of the space program cost-plus FARs made good sense because no one knew how to build a rocket. Having the government absorb the risk was a great way to kick-start the space industry from scratch. But after 50 years it was time to accept that the game had changed. The general term for these new arrangements is public-private partnerships, or PPPs.

NASA's first large PPPs were awarded for ISS cargo resupply services. After a first-round competition award to SpaceX and RocketPlane Kistler in 2006 that only SpaceX survived, NASA added Orbital Sciences (now Northrop Grumman Innovation Systems) in a second round in 2008. SpaceX would build and fly a reusable spacecraft, the *Dragon*, while Orbital would build and fly an expendable module, the *Cygnus*. NASA deliberately chose two suppliers launching on two different rockets to encourage competition and to avoid being reliant on a single means of getting to the ISS, in case one should have a failure. This turned out to be a wise move. The dramatic explosion of the fourth *Cygnus* flight in October 2014 still left one transport system to the ISS.[23] The loss of the seventh *Dragon* ISS flight nine months later, in June 2015, showed that even having two suppliers is risky.[24] The *Cygnus* was back in business in just over a year, and the *Dragon* in just under a year, so the gap was short.[25]

The Commercial Cargo Program has been a great success. In 2014 NASA added Sierra Nevada's *Dream Chaser* to the original duo.[26] *Dream Chaser* looks like a small version of the Space Shuttle. Its wings give it the unique ability to return cargo to Earth from the ISS without putting it through strong g-forces due to rapid deceleration. This is an important consideration for delicate products such as biological materials.

The next step for PPPs is getting astronauts to orbit. The success of Commercial Cargo emboldened NASA to embark on the Commercial Crew PPP in 2010. By 2014, after three rounds of funding to a wider array of companies, NASA selected SpaceX to build *Dragon 2* and Boeing to build *CST-100 Starliner.* The first SpaceX flight with astronauts to the ISS succeeded on 30 May 2020.[27]

As you'd expect, given the look of Tesla cars made by Elon Musk's other big company, the interior of the *Dragon 2* looks like a science fiction movie idea of a spacecraft, with smooth surfaces and touch-screen displays. *Starliner* is more traditional looking, with lots of 1960s-like switches protected by metal guard loops. When I joked about this with one of Boeing's people at a large space trade show, he quickly corrected me. This is not about Boeing being reluctant to change, he said, it's a deliberate design decision. Imagine you are trying to select the right command during ascent while the rocket is firing. It's a very bumpy ride. Are you sure you want the touch screen? Will you really get the right command with your hand shaking badly? That's hard enough to do while riding as a passenger in a car on the highway. It was a clarifying dose of realism, although the astronauts on the first Crew Dragon flight did not seem to have a problem.

The first PPPs have been so successful that NASA has extended the same approach to landing on the Moon.[28] There are many ideas around to extend PPPs even further: commercial space stations to replace the aging ISS, a refueling depot to get bigger payloads to Mars or to the outer Solar System, and a LEO-to-Lunar transport to keep a human exploration base going. Using capitalism to further national goals is a time-honored U.S. approach.

How might a true in-space economy get started? If getting to orbit becomes (relatively) cheap, and continues to get cheaper, then a whole lot of new possibilities for space open up. Where will it begin? What will prime the pump for the space economy? It won't be asteroid mining straight off the bat because selling water in space needs customers. Right now, there aren't any. It's a chicken-and-egg situation. No one will plan on using space-derived resources if they are not available. And no one will make the big investment to provide space resources if there are no customers. Is there a way out? I think so.

First, there will be more humans in space soon. The first wave will take suborbital hops with Virgin Galactic or Blue Origin. These flights will get great views and five minutes or so of weightlessness and they will cross the Kármán-McDowell line of 80 kilometers' altitude that newly defines the edge of space.[29] That will qualify all the passengers for astronaut wings. Later they will go to orbit: for research, for adventure tourism, and to manufacture products in space. The SpaceX and Boeing spacecraft can each carry up to seven people to orbit at once, four more than *Soyuz*, and others look likely to follow.

Thinking 25 years ahead, ULA envisions 1,000 people working in space. The company calls this its Cis-lunar 1000 Vision. (*Cis-lunar* is the hip term for any off-Earth activities within the Earth-Moon system, but not beyond.) ULA will base this vision around a beefier version of its *Centaur* upper stage, called ACES (for advanced cryogenic evolved stage). ACES is meant to be ready for use by the mid-2020s, when it can be launched on the new ULA *Vulcan* rocket. It is designed to be able to refuel with propellant extracted from the Moon. Presumably, asteroid-derived propellant would fit the bill too. George Sowers, then vice president of advanced programs at ULA, said: "I want to buy propellant in space. Once I have a reusable stage and can buy my fuel, then I have the potential to dramatically lower costs to go elsewhere."[30] Part of the Cis-lunar 1000 announcement was ULA's declaration that it would undertake to purchase propellant in LEO for $3,000 per kilogram. By establishing an in-space market for fuel, ULA could be the ones to kick-start the space mining industry. ULA also plans to build a lunar fuel truck it calls XEUS. It is based on ACES. Landing horizontally at a water-mining operation on the Moon, XEUS would then haul the fuel back to LEO, where it could refill the tanks of rockets newly arrived from Earth's surface. A single ACES fuel tank needs 68 tons of fuel. This is all great. ULA is looking at creating a whole system, a true infrastructure, for space development.

There are a few things I find missing from Cis-lunar 1000: what are all those people going to be doing? And who will be paying them to do it? For now, we just have to assume that ULA has a business plan that gives a good chance of a high return on its investment.

One thing they might be doing is going to the Moon. Nothing had landed on the Moon since the final Soviet robotic mission, *Luna 24*, in

1976, until the Chinese *Chang'e 3* went there in 2013. That inaction has all changed. The growing evidence for considerable amounts of ice at the lunar south pole has sparked a lot of interest.[31] Present plans call for a fleet of about 10 robotic landers headed for the Moon in the next few years. A couple have already tried and failed: the Israeli *Beeresheet* and the Indian *Vikram* landers both got very close before their landing rockets cut out early. They were unhappy cases of lithobraking. Included among the upcoming lunar landers are several commercial missions, two from the United States, one from Japan, and one from Europe. NASA is encouraging this trend. It canceled its own Lunar Prospector mission in favor of buying services from commercial companies using the Commercial Cargo PPP approach. Maybe all these lunar missions will lead to more ambitious activities that will provide the customers that Cis-lunar 1000 needs?

With all the ideas for new space activities that are bubbling up, do any of them look as though they will create markets in space for large amounts of water that asteroids could supply? Yes. There are several. None of them are guaranteed, but we may need only one to bear fruit to get asteroid mining going.

Satellite servicing may be about to blossom as an industry. A 2018 report by Northern Sky Research, a space consulting and analysis company, estimates that servicing could bring in $3 billion in revenue over the next decade.[32] That is starting to be usefully big money. The development plan for the successful Northrop Grumman mission extension vehicle is to build a version with 10 backpacks, or mission extension pods, carried by a mother ship. That's a more complicated operation, needing the use of a robotic arm, but having a single spacecraft repair 10 satellites could be a better business case. MacDonald, Dettwiler and Associates (MDA), the Canadian company that built the robotic arms for the Space Shuttle and the ISS, has a plan for a "space infrastructure servicing" vehicle which, unlike the mission extension vehicle, will actually refill the fuel tanks of communications satellites. Other companies, including Astroscale Israel of Tel Aviv, which plans a third approach using ion engines, are also marketing their own satellite-servicing solutions. Astroscale says it has raised $191 million and hopes to launch in 2021.[33] While none of these companies are using hydrogen and oxygen as fuel—and so could not immediately benefit from asteroid water—they will change how communications companies think about their valuable satellites and could open up a market for space-supplied fuel later on.

There is a nearer-term prospect for using asteroid water. Some companies are beginning to make rocket engines that use water as propellant. Without any of the complications of separating hydrogen and oxygen and then cooling them extremely, these engines could be just the market that asteroid miners need. The asteroid-mining company Deep Space Industries was bought by the "green rocket" company Bradford Space because it had this technology. We have already heard about Joel Sercel's Momentus Space, which has developed a clever new version. Meanwhile, a "steam rocket" called The World Is Not Enough, or WINE, is being designed by Honeybee Robotics in collaboration with space-mining expert Phil Metzger of the University of Central Florida.[34] WINE will extract water from an asteroid and use some to refill its propellant tanks in order to move on to the next asteroid. Maybe the same rocket will have uses for satellite refurbishing.

Cheaper, and more reliable, servicing would be easier if refueling was planned into satellite design from the beginning. Daniel Faber's Orbit Fab is developing standards for fuel valves that can be safely built into satellites for this purpose.

A damper on these satellite life-extension schemes is that there has been a decline in orders for the obvious market, GEO communications satellites.[35] It is not clear exactly why this happened. Possibly the telecom companies were waiting to see if the vast new internet constellations of low Earth orbit communications satellites were going to make the traditional GEO communications satellites obsolete. If this decline continues, it will not be good for GEO servicing, yet at the same time another market will emerge. Of the thousands of internet-relay satellites that will inhabit LEO, not all will remain under control. Those dead or misbehaving satellites will be in need of de-orbit services. The orbital garbage trucks required could benefit from being refueled, creating an in-space market for propellant.

A totally new development is the advent of commercial space stations. There are already three serious companies planning to rent out their own private space stations to whoever wants to make use of this resource. Their customers will be not only the major space agencies but also other countries wanting a cheaper route to a space program, and large corporations, foundations, universities, and research organizations, too. They'll be able to do that because of all the companies offering passenger rides to space that we just discussed. This will mean that more people will be in space at any one time, and they will need more supplies. That might open up a market for asteroid water and other useful materials.

Each of the three companies has its own special approach. Bigelow Aerospace is the longest standing.[36] It first flew a small test "station," *Genesis 1*, in 2006, that is still intact and in orbit. The company then attached a larger unit to the ISS in 2016. Bigelow's special innovation is to make expandable airtight modules. All of the original ISS modules had to fit inside its launching rocket's outer skin, its fairing, or the shuttle bay that they came up on. This meant they could be no more than four meters in diameter. The Bigelow modules, instead, grow to about double their starting diameter and several times their length. Bigelow licensed this technology from NASA for its stations. Bigelow already has a smallish unit on ISS—the BEAM (Bigelow Expandable Activity Module) experiment. BEAM has been working well, and NASA now plans to keep it on the ISS instead of ditching it after two years, which was the original idea. NASA's imprimatur for BEAM gave a lot of momentum to Bigelow's plans.

Bigelow is planning to use the same BEAM technology to build its Space Station Alpha. Alpha would be made of two larger versions of BEAM called BA330s. The name refers to the 330 cubic meters of volume the module provides to the astronauts within. Space Station Alpha would have nearly 2/3 the pressurized volume of the ISS itself with just two BA330s, compared to the 14 modules on the ISS. Mike Suffredini, former manager of the ISS, is not convinced. As he told Lee Billings of *Scientific American*, "There's a very big trick in figuring out how you're going to outfit inflatables—where all the plumbing and other systems will go, and how you're going to ensure stale pockets of air don't form inside, since that's something that could asphyxiate a crew. There are all kinds of things that need to be done—and I'm sure they will be—but in the near term I think that's much further away than the time frame we need to fly."[37] Instead, with a rigid module all those systems are built in on the ground, simplifying any (expensive) in-space operations. Unfortunately, Bigelow suspended operations in March 2020 after the COVID-19 pandemic hit.

Suffredini is not an impartial observer. He is a co-founder of a competitor to Bigelow, Axiom Space.[38] Axiom Space was founded in early 2016. It is no start-up full of 20-somethings. CEO Suffredini is a NASA veteran with "30 years human spaceflight experience." Other founders are similarly experienced. There's a lot of tacit knowledge in the space business, as in so many others.[39] There are important details and techniques that never get written down but are passed along by showing and doing. This means that the long experience of Axiom's founders can be a plus, but only if it doesn't prevent new ways of thinking from emerging. Axiom certainly seems plenty innovative.

Axiom's space station design is based on existing modules of the ISS. The modules are rigid metal shells built in Turin, Italy, by the European Thales-Alenia aerospace company. Its three modules on the ISS—*Columbus, Harmony,* and *Tranquility*—are each 9 meters long and 5 meters in diameter. They provide just over 70 percent of the total pressurized volume on the ISS. Because these modules have already flown in space, they are deemed to be "space qualified." That means that space agencies and other customers will feel a lot better about using them.

Axiom's big claim is that its space station will be "at least an order of magnitude cheaper than what NASA spent to build the original ISS."[40] Each Thales-Alenia module on ISS cost about $2 billion, implying no more than $200 million per module. That is a good price. Axiom claimed in 2017 that its whole station, made of a half dozen or so modules, will cost $1.5 billion to $1.8 billion. "By building to ASE industry standards (which came from the auto industry) [Axiom] can use far cheaper components and lets them be replaced easily as 'plug-and-play' modules. SpaceX already uses many of these ASE standards."[41] Changes, even though they cut prices, detract from the modules' "space-qualified" reputation and will cause some heartburn for customers. That's why installation on the ISS is important to let them re-prove the company track record.

Axiom got a huge boost in early 2020 when NASA selected the company to install a new commercial module on the ISS's sole available unused docking port starting in 2024.[42] (Bigelow declined to enter the competition.) Axiom will add more modules over the next five years and will then detach its now self-contained space station from the ISS when it is retired. The company clearly has tourists in mind since it plans to install a windowed space much larger than the "cupola" now on ISS from which to view the Earth. The Axiom cupola and the crew quarters were designed by Philippe Starck, who otherwise designs luxury hotels and yachts.[43]

Ixion is the third space station under development.[44] It will be the product of a new three-way partnership formed in 2017 between Nanoracks, the company selling experiment space on the ISS, the big U.S. rocket company ULA, and Loral Space and Communications. The heavyweight partners, ULA and Loral, used to be exemplars of the "old space" attitude, so their teaming up with a young, small company like Nanoracks is itself a sign of change. Ixion is charting a middle course between Bigelow and Axiom. Like Axiom's, the space station will use a rigid shell for habitat, but instead of a fully outfitted interior it will have the main life-support and station-control systems on the outside. The makers actually have to do this, because their shell is an empty upper-stage rocket used to launch another satellite. Retrofitting a fuel tank is not a new idea. The *Skylab* missions of the Apollo era used the SIV-B upper stage of a *Saturn V* rocket. *Skylab* was a "dry" version, though, meaning that it was launched empty of fuel and with the equipment preinstalled. Ixion will fly "wet," loaded with cryogenic liquid hydrogen and oxygen.

The *Centaur* upper-stage rocket has been a workhorse since 1962, and it is the basis of the new ULA ACES rocket. Even though the *Centaur* stage always gets to orbit when delivering a satellite, ULA has been careful to use the small amount of leftover fuel to slow it down so that it re-enters the atmosphere and burns up. This is good space citizenship, avoiding adding to the swarm of space debris. However, it's a huge waste of high-tech material launched into orbit at great cost. Now ULA wants to stop this senseless waste and recycle the empty fuel tank as something completely different—a space station. From a business point of view, it is a way to get extra revenue from something you are doing anyhow. "We're augmenting what already exists," Mike Johnson, chief designer at Nanoracks, told Loren Grush of The Verge.[45]

The former life of the tank explains the novel approach. The equipment needed for people to live on its inside could not survive being immersed in liquid oxygen and liquid hydrogen at extremely low temperatures. The equipment might also interfere with the steady flow of fuel to the engine. That would not be good. Instead the astronauts will enter the empty fuel tanks through a hatch that is already part of the *Centaur,* bring in some basics, connect up the life-support systems installed on the exterior, and set up any scientific equipment they plan to use. The hydrogen and oxygen tanks are separated only by a thin fiberglass layer. The astronauts will likely remove that and cut through the half-millimeter-thick tank walls to join the two tanks together.[46] That will make for a fairly narrow (3-meter-diameter) but long (12 meters) interior. By adding another hatch and a tube to the front of the tank there can be an airlock. The transformed *Centaur/*Ixion station can then be attached to the ISS or, later, to other modules, or to a visiting crewed spacecraft.

Something not discussed too widely in the press releases is what orbits these stations will be in. The repurposed Ixion *Centaurs* will be in whatever orbit the original satellites were launched into. Half the time they are used to put communications satellites into GEO, so they will be in deep elliptical "transfer" orbits.[47] It's not clear that orbits like that have many uses for a space station. Perhaps they will have enough fuel left to first move them somewhere with a bigger market? Bigelow and Axiom are both beginning by attaching their stations to the ISS. This puts them in the same highly inclined orbit as the ISS, which goes far to the north and south of the Earth. This orbit was chosen for the ISS in order to go over the Russian launch sites to foster international cooperation. That does give good coverage of the populated areas of Earth, which is good for tourism. Who wouldn't want to orbit over their home? On the other hand, for research and most manufacturing purposes, an orbit that goes around the equator is better shielded from cosmic rays, and the rotation of the Earth gives an extra 1,500 kilometers per hour to the spacecraft, which allows larger payloads to be launched to that orbit. It is possible to change the orbit of a space station, but that takes a lot of energy and so a powerful rocket. Most likely new stations would launch to orbits specifically tailored to their customers' needs.

What would be the room rates in a commercial space station? It's hard to predict the market without knowing that. Axiom says they have signed a private astronaut up to stay in their new module for $55 million. That is not entering new low-cost territory. Several years ago, Bigelow Aerospace advertised on its website how much their customers would pay: $25 million to rent 110 cubic meters for 60 days. Using these numbers, Space Station Alpha would generate almost $1 billion per year in revenue. That's assuming customers materialize, of course. The rate is about $200,000/night for each of two astronauts. Amir Blachman, vice president of strategic development at Axiom, claims that there are more than 20 countries that want to fly astronauts.[48] The company predicts that these "Sovereign Astronauts" will comprise a market worth nearly $2 billion annually by 2030. That would be a decent return on investment. Does it add up? By 2030 Axiom plans on having at least two modules in space. Each Axiom module could fly three astronauts at a time. Each of the six astronauts would then have to pay $700,000 a night to generate that revenue. At least you get a room with a view! Although the astronauts will also need a ticket to get to the Axiom station, that sticker price means that a country can still have an astronaut in space for three months a year for $100 million. For countries yearning for a space program that will seem pretty affordable. That's encouraging.

Apart from national-prestige programs, who would use all these for-profit space labs? Axiom says that by 2030 there will be around $3 billion a year in business from customers doing research, supporting deep space exploration, or enjoying adventure tourism; there may also be space factories. Let's start with research.

Physics, biology, and materials science are the fields that look to benefit most from the microgravity of a space station. They are all somewhat speculative, but each one has a basis in small-scale experiments already pioneered on the ISS.

In physics it turns out that microgravity allows better experiments at colder temperatures to be performed than in any Earth-bound lab. When they say "cold" they are not kidding. "Cold" here means just one-thousandth of one-millionth of a degree above absolute zero. In laboratories on Earth strange things happen at these temperatures, thanks to the weirdness of quantum physics. One of these strange things is that a Bose-Einstein condensate can form. A Bose-Einstein condensate is a new state of matter, like liquids, solids, and gases.[49] In a Bose-Einstein condensate the wave nature of atoms becomes obvious, a clear manifestation of the quantum "wave-particle duality." Bose-Einstein condensates have been used to slow light and to build "atom interferometers" that enable extremely accurate measurements of the fundamental constants of physics, because atoms have much smaller wavelengths than light.

Bose-Einstein condensates last much longer in microgravity.[50] Normally gravity is such a weak force that it doesn't matter in most experiments. But at these ultra-low temperatures other forces diminish, leaving gravity as the main force that tends to disrupt this delicate state. A Bose-Einstein condensate can last only a fraction of a second in Earthly conditions. Going to orbit should extend that lifetime a hundred-fold, opening up many new possibilities for experimentation with them. NASA sent the Cold Atom Lab to the ISS on 24 May 2018. Early reports are that it works well.[51] Cold Atom Lab is just a proof of principle experiment that is a pathfinder for much more capable space-based facilities in the future.

Some other, less exotic, materials behave very differently in the ultra-low gravity in orbit. Early experiments with crystal growth showed that they grow more uniformly and larger in microgravity, most likely due to the lack of convection in the liquids they grow in. This has led to hopes for materials with new and useful properties, particularly in semiconductors. At least so far, these experiments have been useful mainly in understanding how crystals grow, and how important influences other than convection are.[52] Many of the results are specific to the type of crystal being grown. Despite over 30 years of microgravity research, it is surprising how much remains to be done. New space stations will help accelerate this work, but it is not easy.

We met granular materials in "Means" when we were considering how to handle a rubble-pile asteroid. But granular physics in microgravity is interesting in its own right. That's because there's no general theory of granular physics. That's abhorrent to physicists, who itch to unify everything into an all-encompassing theory. We've had this unifying itch ever since 1687 when Isaac Newton achieved something close to that, and it has paid off over and over again. Inevitably, then, finding a unified theory of granular materials is the holy grail of the field. Granular physics is a new field, with most of the discoveries happening in the past few decades.[53] Granular materials display a whole lot of weird behavior specific to them.

Some of these depend on gravity; others do not. Being able to explore virtually zero gravity and lower gravity, by means of slowly rotating centrifuges in orbit, could open up a path to getting to that ideal of a unified model.

It's not just physics that could benefit from microgravity. Life sciences have some promising, or at least tantalizing, leads.

Bacteria behave differently in space, as experiments over many years have shown. The assumption is that microgravity is the cause, but it's not obvious why being weightless would have such effects. Suspects have been the lack of convection and the lack of sedimentation in zero gravity, but how this worked was unknown. Luis Zea, of the University of Colorado, Boulder, and his colleagues suggested one possible mechanism in 2016.[54] They pointed out that, lacking convection, the fluid surrounding a cell does not flow in microgravity as it does on Earth. Zea and his colleagues suggest that this altered environment will cause a buildup of acid and a dearth of glucose around a cell. This leads to starvation and creates stress for the cell, which triggers different genes to be expressed.

Cell behavior is also different in space in a more straightforward way. Techshot, a company based in Louisville, Kentucky, wants to use microgravity to let it 3-D print living human organs.[55] Being able to print organs could save the many people who die each year while waiting for an organ transplant. Printing of human ears is already done on Earth, but more complex organs require laying down many different types of cells in complex patterns. Cells are pretty sloppy materials and tend to spread out before you can add the next layer. But that's under Earth's gravity. Techshot thinks that in microgravity this problem will go away. The company has sent a 3-D bio-printer to the ISS to test out the idea. A replacement heart would command a pretty high price. As a human heart weighs only about a third of a kilogram, transport costs to and from space won't be exorbitant. Space could be a sensible place to make them.

That bacteria behave differently in space is very promising for the space economy. Biotech is already as big an industry as space (even including those GPS chips), at $417 billion a year in 2018. It is also growing rapidly. By 2025 it is highly plausible that it will reach $775 billion per year.[56] Biotech is research intensive. If they see a path to profits, the larger biotech companies could each invest tens of millions of dollars into space-based research with little concern. If they find a valuable product they could even begin in-space production. Some drugs have a high enough cost per kilogram that space manufacturing is not going to have a big effect on their price. There are other barriers, though. Few biological scientists are familiar with space and with how its special properties have to be taken into account when designing experiments. Just getting an experiment into orbit involves hardening it to survive the intense vibration during launch.

SpacePharma is a Swiss-Israeli company that aims to make it easy to conduct biological experiments in space. It is building the tools to make it easier for researchers to pursue their ideas both on the ISS and in small free-flying satellites. The hope is that lowering those barriers will elicit a rush of new bio-players in space. SpacePharma could be right. There are already several outfits taking up the challenge. One is Nanobiosym, founded and led by Anita Goel as CEO and chairman. She has sent the "superbug" methicillin-resistant *Staphylococcus aureus,* better known as MRSA, to the ISS to study how to defeat it.[57]

There is a problem. Big Pharma tends not to do basic research itself, leaving that for government or philanthropically funded laboratories. But space agencies have not been experts on drug development. That suggests that a starting point could be a space station devoted to biological research that is funded outside of the traditional space agencies. For example, in 2014 Bigelow announced that renting one-third of one of its BA330 modules would cost $150 million per year.[58] Throw in the tickets up and back for a couple of scientists every couple of months—say, $100 million per year in the early days of $10 million per seat pricing. You have to pay the scientists, of course, and supply them with equipment, but those costs are likely to be small change in comparison, just a few million dollars. That's a total of about $275 million per year. The U.S. National Institutes of Health (NIH) had a $40,000 million per year budget in 2020.[59] Renting lab space on a space station would cost less than 1 percent of the NIH budget. That would seem like a not unreasonable bet to see if doing biotech in space has a good payoff. Those costs are for a continuous research program. It may be smarter to have people working only half-time, or even quarter-time, in space while spending the rest of their time back on the ground understanding their results and thinking up better ways to do the experiments next time around. Then you could start a serious space biotech program for less than $100 million per year.

Why not just use the ISS? There is already research being done on the ISS. Isn't that good enough? The ISS is a U.S. National Laboratory, and NASA has opened up half of its available time and resources to anyone who wins a proposal in open competition. The scientific equipment inside the ISS has become increasingly sophisticated, now including a biomolecule-sequencer, rodent experiment facilities, and protein crystallization experiments in the life sciences and fluid behavior, combustion experiments, and the Cold Atom Lab in the physical sciences. All this activity has led to over 1,500 peer-reviewed research papers based on ISS research. Each new technique added to the ISS expands the range of the investigations that can be done in space. NASA has a great role to play in making more of these research-enabling in-orbit tryouts of new techniques.

Yet there are some serious limitations on what can be done in the ISS. First, the ISS can't handle much more than it does now. The ISS can support up to seven crew members at a time. Typically, though, there are only three. As a result, the ISS astronauts have to be jacks-of-all-trades. Astronauts have to work on maintenance of the ISS itself, sometimes performing space walks (EVAs, or extravehicular activities, in NASA-speak) to do so, which take a long time to prepare for. As a result of all this essential activity, each astronaut gets to spend only about half their time on science experiments.[60] Even then, they have to run experiments from a whole slew of different labs on Earth. The astronauts are extraordinarily talented individuals. Many of them have PhDs, which shows that they know how to do research, as well as pilot's licenses and other impressive qualifications. But they cannot be experts on all these experiments. That makes it hard to get research done on the ISS.

Why not send scientists to the ISS to do their own research full-time? That's not likely to happen. It takes more than two years to train ISS-bound astronauts.[61] They have to learn how to operate equipment from five different agencies and be fluent in Russian to operate the essential Russian systems. Such a long training period for skilled bench scientists would be prohibitive. They have to keep up with progress in their field, or they won't be at the cutting edge. Reading the latest research is time consuming. If you have to spend even one year training to go to space, then you will have fallen badly behind.

Instead, commercial labs will have all their equipment come from a single supplier. And there will be no language training for the current batch of commercial space station providers, as English is the universal language of science. All of this will surely simplify training and will greatly shorten the time it takes to be ready for your flight. Axiom Space expects that about four months of training will be enough. Scientists on the new space stations will have much more time for their research too. The ISS was a first-generation large space station. From it we have learned a huge amount about how to design long-lived structures in space. A newly designed space station will surely incorporate the lessons from the ISS and will use the latest technology. A profit-oriented company like Bigelow or Axiom is motivated to make its space stations last a long time with minimal maintenance to improve its bottom line. Commercial labs will also surely have their own crews on board to deal with the ongoing maintenance, and to take over in case of an emergency. As a result, the scientists who take passenger flights to orbit will be able to put all their time into the research they are good at, without distractions. (Other than the spectacular views of the Earth, that is.) Specialization, as Adam Smith, one of the founders of economics, showed in his classic 1776 book *The Wealth of Nations*, improves productivity.[62] Research on these commercial stations will progress much faster under these favorable conditions.

It is quite reasonable to expect that there will be several commercial space stations within a few years. Each might specialize in one of the research areas like biotech, materials, fundamental physics, or human physiology. Or they may have a mix of users, making it easier to keep rival teams in one technology from doing a little industrial espionage! Each station could support about half a dozen astronauts, so we could see at least a couple of dozen astronauts in space at any one time. That's a big increase over the numbers of the past few decades. They may put up with the Spartan conditions the ISS astronauts live with. But they are likely to require more supplies. Water for radiation shielding, at least of their experiments, could be in high demand. And that could trigger a demand for water from space.

Research may be the first step to a space economy, but to grow to be a significant part of the world economy, there will need to be industrial production in space. Manufacturing in space will be the breakout activity.[63] That's a challenge, as there is nothing right now that could generate bigger profits from being made in space. Just as spreadsheets made personal computers worth having and created a burst of expansion in the industry, we need one thing that is highly profitable to make in space. That would get a new space race going, this time by companies, not superpowers. For something to be better to manufacture in space it has to be worth a lot more than the cost of going to space—even though that cost is set to plummet—and the cost of making it there, with all the complications of vacuum, cold, radiation, and zero gravity.

There is one product that looks like it might meet those criteria pretty soon: ZBLAN. ZBLAN is a "heavy metal" fluoride glass. Its name comes from the five elements—zirconium, barium, lanthanum, aluminum, and sodium (which has the chemical symbol Na)—that are joined with fluorine.[64] ZBLAN has properties that could make it a hundred times better for optical fibers than the ones now being used for high-data-rate internet connections across oceans. Since there is no limit to the number of cat videos we must stream worldwide, there is a virtually infinite market for high-data-rate connections. Ioana Cozmuta and Daniel J. Rasky estimate that a kilogram of ZBLAN can produce 3 to 7 kilometers of optical fiber, which can sell for between $300,000 and $3 million per kilometer. A fairly modest 45-kilogram ZBLAN manufacturing unit could then produce revenues of $0.9 to $21 million.[65] If all that has to be transported is the bulk ZBLAN going up and the spool of optical fiber coming down, then the shipping cost on a *Falcon 9* is only about $130,000, plus packaging (including the *Dragon 2*). A boom is looking likely.

A boom depends on space delivering on the promise of ZBLAN. ZBLAN was discovered in 1974 by the Poulain brothers, Marcel and Michel, and their collaborator Jacques Lucas in Rennes, France.[66] Then, in the 1990s, Dennis Tucker and Gary Workman, two scientists at NASA's Marshall Space Flight Center (MSFC) in Huntsville, Alabama, started investigating how to improve ZBLAN as an optical fiber as part of their long-term basic research.[67] Optical fibers are extruded in a liquid state. When that is done for ZBLAN in Earth's high gravity, crystals form in the fibers. These crystals reflect light going down the fiber and send it out of the sides. The remaining light going down the fiber gets faint rapidly. The very fact that you can see the crystals means they are reflecting light out of the fiber. That's not a good feature when you want to send light thousands of kilometers along ZBLAN fibers. Tucker and Workman tried extruding ZBLAN on short "vomit comet" zero gravity flights. NASA's press release called this process a "sophisticated taffy pull."[68] Scientists wanted to see if low gravity stopped the pesky crystals from forming. The results were startlingly good. Photographs of the fibers made in Earth gravity show lots of shiny crystalline inclusions. In contrast the zero gravity–made fibers show no defects, even though they used the same material and the same equipment. Zero gravity airplane flights give only a few minutes of no gravity, so only a few centimeters of fiber were produced (see figure 9). Showing that such quality can be maintained for many kilometers is quite another thing. That takes industrial-scale research and needs days—or longer—of zero gravity.

That research has started. The company Made in Space (now part of Redwire) of Jacksonville, Florida, has already tested ZBLAN manufacture on a larger scale on board the ISS, where it could get hours or days of zero gravity. When I visited the company's offices, then in the NASA Ames Research Park, in January 2018 I talked with researchers about their ZBLAN tests. They all said that they wouldn't know how good the space-made ZBLAN was until they tested it back in the lab on Earth. On the other hand, they all looked pretty relaxed and happy. Later they told me that they were encouraged enough to fund a second trial. If that goes well, we can be cautiously optimistic. It might well be enough to get venture capitalists to throw money at them to finish the job and fully optimize the manufacturing process. If so, then Made in Space may be leasing whole space stations for bulk manufacture in a few years. In a good sign, the company already has two competitors, Fiber Optics Manufacturing in Space (FOMS) of San Diego and Physical Optics Corporation of Torrance, California.[69]

Large space stations will be wanted if the first small-scale ones prove profitable. Scaling up the size of these stations by more than 10 times may get impractical. Some form of in-space manufacturing will then be needed to enclose large volumes. With an extensive infrastructure of commercial space stations in place there could be a good business in building and supplying these facilities' needs with resources from the asteroids.

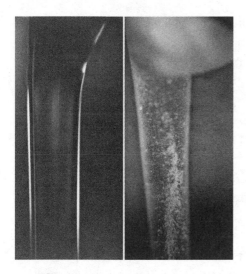

Figure 9: Optical fibers made from ZBLAN material using the same equipment in zero gravity (*left*) and in Earth's gravity (*right*). The zero gravity fiber has virtually no defects that would limit how well it works for transmitting data.

At some point the workers building or using these orbital facilities may find it convenient to live in orbit full-time and go back to Earth only for rest and relaxation. This gradually takes us into the realm of space hotels and, eventually, giant habitats such as those envisaged by Gerard O'Neill and his L5 Society (now folded into the National Space Society).[70] These huge space colonies—or, to use a less baggage-laden term, villages—would have thousands of people living in them. Their building materials would have to come from space. Cue asteroid mining!

There are two really large-scale space endeavors that would almost certainly have to be built using asteroid resources: space-based solar power and space solar geoengineering. Both seem much more fanciful and distant than the uses described so far.

In 1968 Peter Glaser, then working at the management consulting firm Arthur D. Little, proposed putting solar collectors in GEO where they would be in virtually permanent sunshine and then beaming power down to the ground with microwaves. The idea is that there is a lot of sunshine in space that never gets blocked by nighttime or clouds. Also, the ultraviolet light from the Sun that mostly gets absorbed by our atmosphere would be there for the taking. Glaser had credibility, having worked, among other space projects, on the retroreflectors placed on the Moon by the Apollo missions. Glaser's noble hope was that space-based solar power would "lead the world into an era in which an abundance of power could free man from his dependence on fire."[71]

A big reason space-based solar power has not progressed is that the capital cost has always been far too high. The whole solar collector would weigh in at thousands of tons.[72] This is hopelessly uneconomic if you bring the structure from the ground. To launch a thousand tons of steel to a geostationary orbit would cost $10 billion at $10,000 per kilogram. Maybe that can be cut to $1 billion soon, but it still seems uneconomic. Raising the initial capital is itself daunting. As John Hickman of Berry College in Mount Berry, Georgia, says: "Capitalization is a crucial problem for these projects because the total capital investment required is very large and the investment takes a very long time before producing economic returns."[73] Unsurprisingly, then, this idea has not borne fruit.

What, though, if the iron for the structure is almost free? Hickman does agree that if functioning solar power stations were in orbit, they would make an operating profit. If we could bring the tailings of asteroid mining into orbit around the Earth, then that would change the economics radically. Those billions could be cut from the building cost, greatly easing the capitalization problem. Space-based solar power may then become profitable. Otherwise it's a clear case of pie in the sky. This plan depends on the asteroid-mining industry being already up and working, though.

Is there a bridge to space-based solar power via smaller projects? For example, providing power to disaster areas or to remote military operations could be an early small-scale market. Providing power conventionally in these difficult circumstances usually costs far more than normal commercial prices, making space-based solar power more competitive. Not only that, but trucking in the oil needed to run generators can be hazardous and unreliable in both cases. That just might open a window for space-based solar power on a limited scale. The idea is promising enough that the U.S. Navy is conducting tests of the basic technology using the X-37 spaceplane.[74]

The space mining companies themselves may kick off the development of space-based solar power by using it for refining the ore they have extracted up in orbit. To refine the precious metals out of a nickel-iron asteroid large amounts of power, by space standards, are needed. TransAstra, led by Joel Sercel, whom we heard about earlier, is pioneering solar power for refining ore by testing its designs on the ground using solar concentrators. Building large ore-processing facilities for the precious metal using the tailings of early mining expeditions may become a way to increase profits. From there it would be a modest step to larger facilities.

There is another objection to space-based solar power. The intensity of the beam coming down to the antenna on the ground has to be a lot higher than that of sunshine, otherwise why not just build solar panels covering the same area? Glaser originally suggested a beam 10 times higher than sunlight—1 watt per square centimeter. This intensity raises concerns. At that level living tissue in the beam would soon be damaged, Glaser said. While people and airplanes can be routed around the beam, birds could not. This feature makes many unsure about the ethics of space-based solar power. Deliberate use of the beam as a weapon in a more focused form is also possible, either by the operators, or by others hacking into the control system. That could be a killer—for the idea, as well as for people.

There is a new idea for how we could use beamed microwave power. There is no really good way for us to cross the oceans without emitting large amounts of carbon dioxide. Airplanes and ocean liners are both great emitters of greenhouse gases. Yet we have to stop emitting them if the planet is not to warm too much for our own good. Shipping could, in principle, be solved by using nuclear-powered ships. Certainly, nuclear-powered aircraft carriers have a long and safe history.[75] Airplanes are much harder. Electric airplanes powered with batteries or fuel cells don't look likely to get the size or range needed to replace long-distance airliners. They just don't store energy as densely as aviation fuel.[76] Are we just going to stop transoceanic voyaging?

Microwaving power to airplanes in flight just might be a way out. If we beam power to the airplanes, then they wouldn't have to carry any fuel at all. (Well, maybe just enough to cope if the beam were to falter.) Justin Lewis-Weber, CEO of Empower Earth, a company looking to commercialize long-range radiative power transmission, has suggested beaming the power up from the ground, which works for crossing a continent. But how do we deal with the wide oceans? To a space enthusiast the answer is obvious. Beam the power *down* from space. This would need at least several big antennas in orbit. The antennas need to be large to make the microwave beams small enough that an airplane can pick up most of their power. That's for efficiency, but also because not too much should reach the ground for safety reasons. Using lasers instead of microwaves would make the receiving dishes much smaller and more practical. Safety could be an issue for accidental—or malicious—pointing. Planes fly above almost all clouds, so the beam—microwave or laser—won't be stopped by the water vapor that both are sensitive to. They fly above the birds too, so none will get microwaved. Construction of those antennas would be a large project. There are companies working on ways to do that, with communications in mind. Made in Space is one; Tethers Unlimited's Firmamentum division is another.[77]

Space solar geoengineering is an even bigger idea emerging for space manufacturing.[78] We seem to be failing the challenge of climate change. The prospect that we can reduce our carbon emissions in time to stop the world warming by more than 1.5 degrees Celsius is looking remote. That impending failure has led a small band of folks to suggest we give ourselves some breathing room to get the job done. Their suggestion is to find a way to cut the amount of sunlight reaching the ground. The amount of reduction needed is just 1.8 percent, a surprisingly precise figure. That could buy us several decades to undo the carbon dioxide pollution of our air. Modifying the amount of sunlight reaching the ground worldwide is one part of the ambitious, some would say grandiose, field of geoengineering. It is called solar geoengineering.

The quickest and cheapest way to make a sunlight veil that would intercept enough sunlight is to inject sulphates into the stratosphere. We know that this will reduce global temperatures because that's what happens during giant volcanic eruptions. In 1990 Mount Pinatubo in the Philippines erupted, putting 20 million tons of sulfur dioxide into the stratosphere, where it reacted to make a haze of very small droplets of sulfuric acid. These droplets reflected sunlight back into space and reduced global temperatures by half a degree Celsius for the two years that the haze stayed in the stratosphere. We could reproduce this effect deliberately using a fleet of high-flying planes to inject the sulphates where they are most effective. This would be much more efficient than an eruption, and so far less material would be needed than Mount Pinatubo provided.

Many of us instinctively react against the idea of any geoengineering. Do we really understand enough about what we are doing to be sure that there won't be unexpected consequences? And might these consequences be even more dire than the problem we're trying to solve? You'll be relieved to know that geoengineers in general agree with this caution. That's why they want to make small-scale experiments now to find the "gotchas" in all the proposed techniques, so we'll be ready if we find we really need them in a hurry. There's very little funding going into such experiments just now, though the amount is increasing. Still, messing about with the air we breathe has got to be worrying at some level.

That's why some geoengineers have promoted the idea of blocking the sunlight before it ever gets to the atmosphere, by putting up a veil in space. I had the fun experience of going to a one-day workshop at Harvard put on by David Keith, a pioneer of geoengineering. Maybe I am just easily convinced, but I walked in a skeptic and came away much more positive about geoengineering.

I was invited because, even if the screen were wafer thin (say 1/100 of a millimeter thick), it would still take at least a million tons of material to construct, as it has to be the same diameter as the Earth. Launching that much mass from the Earth would cost hundreds of billions of dollars, even at low predictions for launch prices. But a million tons is the mass of just one fairly small asteroid, maybe 100 meters across. It takes 100 times less energy to get an asteroid to the sunward location where a screen would be constructed than it does to bring the tonnage up from the Earth. Using asteroid iron might just make this idea a doable proposition. If we build the capability to save the planet this way, then we will also have built a large in-space industry that can later be turned to all the other uses we've been discussing. Perhaps saving the planet with space solar geoengineering will be the chicken in our chicken-and-egg problem of getting a space economy going.

Back to the immediate future. Why just have scientists in orbit? Why shouldn't anyone, providing they have a large pile of cash, buy a ticket to space just for the fun of it? It would address the problem that billionaires have, expressed so well by the baroness in *The Sound of Music:* "Where to go on our honeymoon? Now, that's a real problem. A trip around the world would be lovely. And then I said: 'Oh, Elsa, there must be someplace better to go.' "[79] Soon there may very well be. Many space fans think tourism will give the first big payday to the new space economy.

Space tourism has already begun. Space Adventures is a company that sent seven people to the ISS as passengers on the Russian *Soyuz* spacecraft between 2001 and 2009.[80] Space Adventures is a more honest name than Space Tourism. It clearly admits that the level of risk will be high, and the level of comfort minimal. Space Adventures' customers had to train exactly the same way as astronauts. This will surely continue to be true at first.

The first space tourists to fly in anything other than the Russian *Soyuz* vehicle won't get all the way to orbit. They'll be making suborbital hops with Virgin Galactic and Blue Origin, rocketing up over 80 kilometers, where the sky turns black. For five minutes they will be weightless and can view the curve of the Earth below. Then their spaceship will fall back down.

The idea for this type of tourism started to become real when serial airplane innovator Burt Rutan's company, Scaled Composites, built a vehicle that won the $10 million Ansari XPRIZE in 2004.[81] To win it had to fly the same vehicle to space within two weeks. The prize givers decreed that flying above the Kármán line at 100 kilometers' altitude qualified as "space"; it didn't have to go into orbit. That's important because getting up to 100 kilometers takes 50 times less energy than to get to orbit. You only need to reach speeds of 3,500 kilometers per hour (Mach 3.5) rather than the 28,000 kilometers per hour needed to stay in orbit. Rutan's design featured a clever folding tail section to slow the vehicle down in the upper atmosphere that then folds back into a normal glider shape at about normal aircraft heights (10 or 20 kilometers up). The space plane then flies back down to a standard runway. Rutan's *SpaceShipOne* now hangs in the Smithsonian's Air and Space Museum in Washington, DC.

Billionaire Richard Branson immediately jumped in after this winning flight to partner with Scaled Composites, forming a new company called, with classic Branson immodesty, Virgin Galactic. It would offer rides to space to the general public in an upgraded vehicle, *SpaceShipTwo*, to carry six passengers and two pilots. The new ride would feature large portholes for the tourists to view the Earth and would give them those few minutes of weightlessness at the top of the arc of their flight. Its style and panache made Virgin Galactic the face of space tourism for a decade. The company managed to sell several hundred tickets at $250,000 each.[82] That certainly shows there is enough demand for at least a couple of years of weekly flights. Things didn't go as quickly for Virgin Galactic as it had hoped, though. A disastrous test flight in October 2014 ended in the disintegration of its test vehicle and the death of one of the astronauts. This added to the company's delays and has allowed new entrants to the game.

Blue Origin, amply funded by mega-billionaire Jeff Bezos, has now repeatedly landed its crew capsule and its *New Shepard* single-stage rocket back at its West Texas launch site. Blue Origin is offering the same five minutes of weightlessness as Virgin Galactic. Unlike Virgin Galactic, there will be no Blue Origin crew on these trips. In fact, the video tour of the (so far nameless) capsule doesn't show any controls at all![83] The first flights with people on board will be with volunteers from among Blue Origin employees.

Blue Origin has a strongly customer-oriented approach, as might be expected from a company founded by the man who started Amazon. This is clear from that first video. It touted the *New Shepard* capsule as having "the largest windows in space," implicitly inviting comparison with the smaller portholes of Virgin Galactic's *SpaceShipTwo*. They also say they will take passengers "past the Kármán line—the internationally recognized boundary of space" at 100 kilometers' altitude. This is a dig at Virgin Galactic, which will only get above 80 kilometers. This makes Virgin Galactic keen supporters of the newer Kármán-McDowell line definition of where space begins which, fortunately for them, is at 80 kilometers.[84] Advertising the final frontier may seem gauche, but it will pay the bills. And paying the bills to make a profit will grow the industry and lead to the next step: orbital passenger flights.

Going to orbit will be a much more impressive adventure than a suborbital hop. Just one orbit will get you 90 minutes of zero gravity instead of 5, and as you circle the Earth you will have astounding views of the continents, mountains, and rivers drifting by below you. As the George Clooney character in the movie *Gravity* said as he floated out in space: "Wow, you should see the sun on the Ganges. It's amazing."[85] I'm sure he's right.

Suborbital flights will quickly come to seem like the poor man's space experience. (Well, not *so* poor.) Blue Origin is clear that its suborbital business is a way to practice and gather revenue on the way to orbital flights. The name of its first rocket, *New Shepard*, was a bit of a mystery at first. Was it some kind of poorly spelled biblical reference? All became clear in September 2016 when the company announced its next rocket, the one that would go to orbit. It was called *New Glenn*. Space buffs immediately knew that Blue Origin is naming rockets after NASA astronauts who made the first key steps: Alan Shepard made the first U.S. suborbital flight and John Glenn made the first U.S. orbital flight. In fact, Blue Origin is also talking about a Moon rocket, much larger than *New Glenn*, that it calls *New Armstrong*. Neil Armstrong, of course, was the first person to step onto the Moon.

Blue Origin has competition. At least four other companies have been working on passenger services to orbit. Three of them we've already talked about. SpaceX and Boeing have their NASA-funded vehicles to ferry crew to the ISS, though problems with both systems made for delays. The wait is over though. On 27 May 2020 the Demo-2 flight to the ISS of two NASA astronauts aboard a Crew Dragon set the way for regular trips to orbit on a commercial vehicle. Flights for other customers are planned to follow. Both companies have announced that they will sell tickets to tourists and others. Sierra Nevada, which already has a NASA contract to ferry supplies to and from the ISS, is also planning to be in the business of sending people to orbit. Its *Dream Chaser* spacecraft, unlike *Starliner* and *Dragon 2*, can land back on normal airport runways. This version of the *Dream Chaser* is a downscaling of the company's original personnel-carrying vehicle. Passenger transport remains a company objective.[86]

Least splashy was XCOR. Its one-crew, one-passenger *Lynx* spacecraft was to begin, like Virgin Galactic, as a suborbital vehicle, but was designed to scale up to an orbital capability. Unfortunately, XCOR went bankrupt after its only customer withdrew.[87] That's how capitalism crumbles. Hopefully, XCOR's intellectual property will be picked up by another company from Build a Plane, which purchased it all, as the concept looked fine.[88]

If there is a strong demand for these orbital joy rides, there will be many companies competing for passengers by the early 2020s. XCOR was aiming for a ticket price of $1 million to orbit, which appears quite plausible. That is a price point that surely opens up to many more customers beyond state agencies and giant corporations. In value for money that is a big win over a suborbital hop. For 4 times the price you get at least 20 times as long being weightless, and views all around the globe. Then space tourism will really begin.

Short orbital flights won't themselves require any asteroid resources. But they will pave the way for longer stays in orbit by tourists. Space Adventures is already selling tickets for a five-day orbital flight on the Crew Dragon in the $10 million to $20 million range, and Axiom is already selling 10-day stays in its ISS module, albeit for $55 million.[89] That will lead to a need for orbiting "hotels," and those will need so much more in the way of supplies that getting some from asteroids could start to make economic sense.

At first conditions in the commercial space stations will be Spartan. Maybe showing off your astronaut *Right Stuff* toughness will be part of the appeal? Even though you paid $10 million or more, there is no shower on the ISS, though the much earlier *Skylab* did have one.[90] The suction system in the toilet on the ISS has been pretty awkward to use and unreliable. Fortunately, a major new "universal waste management system" should be a major upgrade.[91] Commercial space stations will have strong incentives to solve these problems.

One good thing that will help adventure tourists adapt is that there are now ways to treat the motion sickness that comes from being weightless. Apparently, the first few minutes of weightlessness are not a problem, so suborbital tourists won't be overly troubled. But after a while your brain reacts to the inconsistent signals coming from your eyes and your vestibular system, the mechanism in your ears that gives you your sense of balance and orientation. Then you throw up. It is a very common problem. The second person to orbit the Earth, Soviet cosmonaut Gherman Titov, became the first human to vomit in space in 1961. Space sickness seems to be worse in larger spaces, so space hotels might be particularly bad. The symptoms vary a lot from person to person, and generally wear off after a few days. The drugs that suppress this unwelcome side effect of space will surely help the tourist trade.

Astronauts on the ISS have their own small sleeping cabins. As they are weightless, they don't have a bed with a mattress, just a place to attach themselves to a padded board in a sleeping bag. But the ISS is a noisy place. NASA would like to keep the noise down to 70 decibels (formally dBA), about the loudness of a dishwasher or shower, but does not seem to have achieved that yet. The walls of the sleeping cabins are not designed to cut down the noise. Tibor Balint and Chang Hee Lee, designers from JPL and the Royal College of Art in London, respectively, have taken up the challenge to create a more comfortable and appealing sleeping compartment.[92] They started with the idea of a pillow and what it should do in this novel environment. It should, of course, shield your head against hitting the cabin bulkhead. But it could also have built-in noise reduction and eyeshades. That would be easier, and lighter, than soundproofing the walls.

Design in space has been very much "form follows function": unless something is technologically required it really doesn't get considered. The astronauts involved in the Apollo-era *Skylab* space station were "somewhat disdainful of the attention given to such amenities as interior color schemes" by the famed industrial designer Raymond Loewy and others.[93] Not much has changed. The few parts of the walls of the ISS that are not covered with storage are gray.[94] In 2017 Janne Robberstad, a Norwegian artist who works with ESA, was working with schoolchildren on one of her Global Science Operas, this one set in a lunar base. When her students began designing the sets she asked them what color the walls should be. The kids were puzzled. "Don't they have to be gray?" they said. "Well, no. It's up to you," Janne told them. Once space tourism has gotten past the early "tough astronaut" phase, they'll be any color you fancy. Already the cabins that Philippe Starck is designing for Axiom Space look more like a five-star hotel room than a hostel.

Creature comforts for space travelers are on their way too. The space espresso era has already begun. In 2015 Italian astronaut Samantha Cristoforetti was the first to have an espresso in space. Lavazza, the Italian coffee manufacturer, partnered with engineering firm Argotec to build the ISSpresso (their pun) barista machine for the Italian space agency.[95] Research runs on caffeine, so a boost in ISS productivity should result! That's work taken care of. What about play? Surely, you'll want to celebrate becoming a space-farer? No problem—space champagne has arrived in the form of Mumm's Grand Cordon Stellar, released in 2018, initially for zero gravity flights aboard an airplane. Apparently, it tastes even better in space than on the ground. According to Mumm cellar master Didier Mariotti, who sampled some Grand Cordon Stellar during an Air Zero G flight, "Because of zero gravity, the liquid instantly coats the entire inside of the mouth, magnifying the taste sensations. There's less fizziness and more roundness and generosity, enabling the wine to express itself fully."[96] Now if only they can fix the toilets.

For-profit space station hotels will emerge with enhanced features. Some may have spinning parts to provide a little pseudo-gravity from centrifugal force. Having just enough to make the showers and toilets work more like they do on Earth would be a good start. Improved hygiene would be quite the sales hook for tourists when they compare which space hotel to patronize. These spinning sections don't have to be the complete wheels common in science fiction movies. The first ones might be more like bolos—two large cans with smaller tubular access ladders from the main station. They would have to be carefully placed symmetrically to keep the station balanced. Learning to use their gangways would take some practice. The obvious thing would be to float in headfirst from the main space station but, as you accelerate, you'll soon realize this is not a good idea. Feetfirst is the only way to go.

What do you do when you get to your space hotel? Gazing at the Earth going by and trying to spot your hometown as well as iconic coastlines, rivers, and mountains (not to mention watching the Sun on the Ganges) will be enough at first. After a few days of rapt viewing of the Earth, though, you'll probably want to be doing something more active. My bet is that it will be the fun of flying around in weightlessness that's special. The astronaut Ed Lu decided that spending his free time on the ISS reading paperbacks was a wasted opportunity. Instead he spent all the time he could learning to fly around skillfully.[97] He won't be the only one who'll get a kick out of doing that, although be warned: it's not easy.

As more tourists get to fly, they'll inevitably start having races and playing tag and the like, bouncing off the carefully protected walls. Eventually new sports that can be played only in space will arise. Makoto Arai of Dentsu, Inc. in Japan is developing space sports.[98] He points out that for astronauts on long-duration stints in space it is boring to spend several hours a day on exercise machines. How much more interesting it would be if they could play sports instead. Arai has convinced the International Olympic Committee to sponsor sports events in space. After that he envisages "bird-human" races on the Moon, where the reduced gravity should make it possible for us to fly like birds by flapping wings—inside large, air-filled domes, of course.

Quidditch could be one of these space sports. J. K. Rowling's fiendishly complicated game for witches and wizards from her Harry Potter series seems ready-made for space. Quidditch players zoom high above the playing field on broomsticks trying to get the quaffle though one of the hoops while avoiding dangerous bludgers and chasing the ultimate prize, the Golden Snitch. Swooping around on "broomsticks" powered by compressed air, surrounded by similarly self-propelled bludgers and snitches, Quidditch players in space could be just like true wizards and witches. Unfortunately, a Quidditch field is big: "an elongated oval 500 feet long and 180 feet wide," Rowling writes.[99] That's rather too large. The largest space station planned so far is Bigelow Aerospace's *Olympus* module. It is huge, but still only a tenth the size of a Quidditch pitch. So true Quidditch games will have to wait. Maybe we can start with peewee Quidditch? Once the demand is there, then surely the supply will follow?

For more culturally minded space tourists, microgravity could usher in a revolution in the arts. Yusaku Maezawa, a Japanese billionaire, has put down a hefty deposit on a trip around the Moon in the SpaceX *Starship*.[100] He plans to take half a dozen artists with him. What special things might they do? Imagine Cirque du Soleil in space. Dancer, circus artist, and physicist Adam Dipert is studying zero gravity dance moves. He built a simple computer model of a human body to show how it would spin in zero gravity, depending on where the legs and arms were positioned. After finding some good tricks in theory, he went on several parabolic flights to test them out in practice, with promising results. He's hoping to get an orbital flight someday to develop them into a real choreography for space. Other arts will be transformed too. MIT's Media Lab has a Space Exploration Initiative designed to explore a huge range of activities in space, including the arts.[101] The vision of Ariel Ekblaw, the founder and lead of the MIT initiative, is that "space will be hackable. Space will be playful."

What about safety? Immediately after praising the great view from space in the movie *Gravity*, the George Clooney character—spoiler alert!—dies. Won't the whole space tourism industry shut down as soon as there's a fatal accident? After all, when the *Challenger* blew up on takeoff in 1986,

NASA stopped all shuttle flights for two years. Then, after the *Columbia* broke up on reentry in 2003, NASA again stopped all shuttle flights, this time for three years. Expert panels and congressional inquiries followed, with massive reports being issued. Why would space tourism be any different?

It might not be, but that doesn't fit with the way things have played out in other voluntary and dangerous activities. Occasionally tourists have been eaten by lions while on safari, but people still flock to the game parks.[102] Extreme sports can have remarkably high fatality rates. The most dangerous seems to be BASE jumping, in which people jump off high places wearing suits with web-like wings between their arms and legs. (BASE stands for the four types of high places they jump off: buildings, antennas, spans [bridges], and earth [cliffs].) The death rate for BASE jumpers is about 1 in 60.[103] That's about the same risk as flying in the Space Shuttle. Some people can tolerate high risk. How many of these individuals, though, are also rich enough to be tourist astronauts? How safe would space tourism have to be for those of us who are less risk-tolerant to be willing to fly to space?

The history of steamships and jet planes suggests that a few unfortunate deaths won't have much of a dampening effect. Steamships enabled transoceanic travel, even for the poor, in the mid-nineteenth century. Think of the 2 million impoverished Irish who came to America to escape the potato famine. Yet the steam engines on those ships often used to blow up.[104] Despite the obvious danger, people continued to buy tickets and the passenger liner industry grew. Eventually, improved technology and strong safety regulations—almost always prompted by major disasters—brought the risk down, though not to zero. Refining the system never stops. Railroads were no different. It was only after a 1996 train crash in Maryland that the United States imposed uniform rules on passenger railroad cars. The same happened with jet engines in the mid-twentieth century. The first operational passenger jet plane was the de Havilland Comet in 1954. You likely haven't heard of it because it used to drop out of the sky rather often. It was quickly taken out of service. The primary cause was found to be metal fatigue.[105] But jet travel didn't stop. Instead the U.S. Boeing 707 jet came along a few years later, was much safer, and swept the market. Again, as with steamships, the 707 was not perfect, but it crossed a safety threshold that people could accept.[106] In neither case did an accident, or even many accidents, shut down the new industry.

When anyone can fly in space, who, then, is an astronaut? Just over 500 people have flown in space from 1961 (Yuri Gagarin, at the beginning of the Space Age) to 2019. The first few Americans were given ticker-tape parades down the avenues of New York. It was a big deal in the 1960s. Even now, astronauts are treated like heroes, although on a smaller scale. Will all the passengers on Blue Origin and Virgin Galactic suborbital "hops" be astronauts? If each company were to fly weekly, with six people at a time, then within a year there will be more tourist-astronauts than all of their astronaut predecessors. This feels to me like it devalues the title of astronaut. When the X-15 rocket-plane pilots and the *Mercury* astronauts, Alan Shepard and Gus Grissom, did their five minutes of weightlessness in suborbital flights, their feats were truly heroic, with a real chance of death. Should we try to reserve the term *astronaut* for bolder adventurers than the new tourists?

Descriptions do change in their power. "World traveler" used to be exotic too, but with college students going across the globe in their gap years, the term now just reveals that their parents have some money. "Explorer" used to be a real job title, at least in the West. Most of the places explorers went already had people living there, of course. But there was a sense in which explorers were integrating knowledge of the world into a single whole, and they certainly took huge risks. But the job of explorer is gone. We have that integrated picture now, as Google Earth testifies. For a while, people could still be the first to do something: climb Everest, reach the deepest places in the oceans. Then they had to get more extreme: climb Everest without oxygen, join the 7 Summits Club by climbing the highest peaks on each continent within a year. As the goals became more elaborate, they became less exploratory, less adventurous, even as they remained strenuous.

Should we redefine "astronaut" so it is granted only to those who go beyond what has been done before? Or should we invent a new term for those who truly explore space? Perhaps we should resurrect "explorer" and use it for those astronauts who really go beyond the frontiers of the known—the first few to do something bold and new? Calling suborbital tourists astronauts doesn't seem to have much support among space enthusiasts. In December 2018 ParabolicArc.com asked in a tweet, "What do we call passengers who fly to 80 km, which requires a small percentage [3 percent] of the energy needed to get to orbit?"[107] The choices were given as: (1) astronauts, (2) spaceflight participants, (3) CG [center of gravity] ballast, and (4) the 3 percenters. The votes were split quite evenly among the last three choices, but only 14 percent voted for astronaut. Space Adventures customers have tended to prefer to be known by other terms, including "private astronaut" and "private space explorer." For now, "spaceflight participant" is the term used by both NASA and the Russian Federal Space Agency.

There's also a legal issue in changing the definition of astronaut. The United Nations Rescue Agreement provides for "all possible steps" to be taken to rescue astronauts in peril.[108] But it clearly refers to astronauts rather than, say, spaceflight participants. If we redefine astronaut to be more exclusive, what about space tourists in peril?

All of the examples in this chapter show that there is great *potential* for doing business in space. Nothing is ready to burst onto the scene just yet. Within a decade some of these initiatives seem likely to succeed. There's no guarantee of that, of course, but the possibility is there. Once these activities cross a threshold in size, they will generate a need for resources from space. Water brought in bulk from space for tourists or industry could then become cheaper than bringing it from the ground. As space resources become available, other uses will become feasible. After perhaps a brief interlude of using lunar materials, which are probably easier to extract but not so abundant, the mining of asteroids will become the way to satisfy these new markets.

9

Making Space Safe for Capitalism

How do we get to the jumping-off point where profits start to flow from space ventures, including asteroid mining, so they become self-propelled and grow of their own accord with no need to appeal to taxpayers and legislators to get funding? Will pure free enterprise be enough? Or will governments have a role? There are already up-front government investments in technology to buy down risk. That won't be all though. There are legal and regulatory hurdles to clear before the risk for investors becomes acceptable. How do we make space safe for capitalism?

Why not leave space to the private sector? There are well-known benefits to private enterprise: the ability to deliver speedy results, a strong incentive to contain costs, and the energizing effect of competition on innovation. On the other hand, commercial ventures tend to have short-term horizons because they have to pay the bills and keep their investors happy. The opening up of the American West to European-descended settlers is often invoked as an analogy for the "space frontier." Often, though, that Wild West is portrayed as a lawless place where the government's writ did not run, and bold pioneers struggled alone against the odds. That is not exactly how it went down, even ignoring the repugnant treatment of the people already living there. Brave and bold as they surely were, the miners and farmers setting off from Independence, Missouri, in their Conestoga wagons were not setting off into unknown lands. It's not so widely appreciated, but the U.S. government had invested heavily in mapping the West and its resources, both water and mineral. Public investment can stretch over decades, sinking the costs and buying down risk for the capitalists who follow. William Goetzmann explained this in his 1966 Pulitzer Prize–winning book *Exploration & Empire*.[1] Goetzmann's book traces the dozens of government-funded expeditions that mapped the American West, pinpointing its resources in ever-greater detail, from Lewis and Clark's exploration of the vast and newly acquired Louisiana Territory in 1804–6 up until the 1880s.

The example of John Jacob Astor suggests why private enterprise has limitations. Back in 1810, only four years after Lewis and Clark had returned from their exploration, Astor, America's first self-made millionaire, decided to extend his fur-trading empire by setting up a trading port at the mouth of the Columbia River (now in Washington State). Goetzmann's book tells the tale. Astor was trying to beat the Hudson's Bay Company to the punch. That part of the West Coast had unclear sovereignty; both the United States and Britain wanted to control it. Astor invested heavily in his Pacific Fur Company. He had a ship sail around Cape Horn to the mouth of the Columbia, where he set up Fort Astoria as his base in 1811. He also paid for an overland expedition from St. Louis that got to Fort Astoria a year later. It was a sensible plan; there were plenty of beaver to be hunted for their pelts. But for some reason the expedition members weren't able to find them, so the disappointed partners were already planning to leave when the HMS *Raccoon*, a British Navy ship, arrived in 1812 and took possession of the fort without a fight. This story illustrates how a private company can act fast, but also how its attention span is short if there is no immediate profit to be had. Private investment like Astor's has to be short term because companies have to stay in business.

Marianna Mazzucato is an economist now at University College, London. In 2013, when she was at the University of Sussex in the United Kingdom (where long ago I got a master's degree in astronomy), she published her book *The Entrepreneurial State*.[2] In it she shows how long-term government investment is behind most of today's high tech. Mazzucato uses the Apple iPhone as an example. From hardware in the iPhone—CPUs, LCD displays, lithium-ion batteries—to the infrastructure it relies on to make it great—the internet and the World Wide Web, GPS, cell-phone technology—to software—such as the AI for SIRI—all were developed with government funding. Apple's genius lay in integrating all these technologies into an indispensable and beautiful whole. The same story, Mazzucato shows, applies to pharmaceuticals, biotech, wind and solar power, and more. Long-term basic investments need government support precisely because they can't generate profits until they are mature. In this area government has a good track record.

These examples show how governments can take the long view and absorb costs that would be intolerable to a private enterprise. They can bring down the risks to a level that private investors can cope with. Governments can invest directly in companies too. Unlike angel investors or venture capitalists, governments do not typically take an equity stake in early-stage companies in return for supporting them. This allows the founders to develop their products without diluting their stake in their companies too soon, which helps keep them motivated. Once a profit-making industry is in place, governments can step aside from the technologies and let capitalism do its job. Even then governments have important roles to play in the legal, regulatory, and diplomatic side of commerce.

Governments, if they go about it the right way, can be the pump primers for space commerce. This will work to their benefit. The new space businesses will pay taxes. Their own space programs will gain too as space exploration will be able to make use of the greater and far cheaper capabilities enabled by ever-larger private companies. NASA's roughly $20 billion budget may seem large now, but that's because it supports only pure science and space spectaculars. If instead we think of it as an infrastructure investment, designed to open up great new economic possibilities, then it is actually quite modest. As a result, government activity in space can be expected to grow as space activities get cheaper. The true exploration of the Solar System will kick into a higher gear once economic gain gets involved.

The United States, Russia, Europe (via the European Space Agency), Japan, India, and China have all invested significant funds in space over seven decades. Government actions created the whole possibility of space as a field of human endeavor, including commercial enterprise. They developed the technology needed, mapped the Moon, and found many near-Earth asteroids. Their motives may not have been primarily to develop an industry in space, but they helped. The NewSpace movement may inspire them to do more. They could make the enabling of profits from space enterprise the centerpiece of their twenty-first-century space policies, as Luxembourg does today. NASA's primary directive, from President Reagan's administration in 1985, was the updating of the original 1958 Space Act to include the mandate to "seek and encourage, to the maximum extent possible, the fullest commercial use of space." The government could accomplish this in many ways: by pushing technology, by mapping out the territory, by providing a convenient asteroid to practice on, and by being the anchor customer for commercial space stations and other vehicles. The Commercial Cargo and Commercial Crew PPPs have started NASA down this path. Commercial use of space is not yet a NASA directorate, one of the top-level organizational units. But it could be. Doug Loverro, shortly before stepping down as NASA associate administrator for human exploration and operations, announced in April 2020 that his directorate is reorganizing to create a division devoted to the commercialization of space.[3]

Governments, however, are also notoriously prone to being slow, wasteful, and averse to innovation. Governments, naturally, also choose approaches based on political pressures. The wrong type of public investment can stifle rather than encourage private initiatives.

Philanthropy that values social benefit and invests generationally offers a third way forward. Philanthropists may buy down risk for capitalists. Historian Alexander MacDonald, the chief economist at NASA, argues in his book *The Long Space Age* that it was philanthropy that began space exploration in the United States by funding the great astronomical observatories of the early twentieth century, such as the 200-inch telescope on Mount Palomar.[4] The B612 Foundation initiative to survey for asteroids that are hazardous to Earth is another example. B612 undertook its Sentinel mission to find all the potentially hazardous asteroids because NASA had not taken it on, despite the George E. Brown congressional mandate to do so.[5] Its work was directly relevant to asteroid resources, as it would have found many nonthreatening but easy-to-reach asteroids too. Philanthropy can seed new areas that are too risky for private investment and not yet a priority for government funding. In this way they can fill gaps in programs. Philanthropic funds, though, are relatively limited in size compared to private or public funding and so tend to be limited in scope and may not be as enduring as governments in their purpose. B612 eventually had to give up on its ambitions. NASA now has definite plans for such a mission.

The right balance between public, private, and philanthropic methods is not obvious. It is always affected by deeply held ideological beliefs. Even if this were not the case, the most effective mixture for enhancing human presence in space, our assumed goal here, is not likely to be constant over time. Getting the best from all three is an endless balancing act.

What could governments do? The economic development of asteroids and other space resources will take a wide range of skills. There are five main areas—technology, exploration, policy, regulation, and diplomacy—and the last three are uniquely the role of government. The first two could be undertaken by commercial or philanthropic organizations, but when a task offers only long-term rewards it is generally left to governments. There is much to be done in all five areas.

First, technology. Every part of space development, from commercial space stations to asteroid mining, could benefit from government investment. NASA is already assisting in some of these, both by its own research and by providing grants and contracts to a host of (mainly small) U.S. companies. NASA expertise, the International Space Station, and other human activities in low Earth orbit can be used to great effect. Sometimes that can mean the space agency directly buying services, such as space stations or lunar landers. More often it is developing techniques, for example, bringing a large piece of an asteroid nearby where we can practice all the risky maneuvers needed to put a mining machine on the asteroid, excavate the rock, and extract the ore. None of this has been done in the microgravity and vacuum of space before. It's best not to try it for the first time when your mining craft is far away. Even if it is eventually done robotically, having people nearby is likely to be a great help early on in working out how do it right, so human spaceflight is part of the deal. Similarly, faster transport—using souped-up ion engines or nuclear rockets—has to be developed. Government could be the anchor customer to get these tools into production fast. To some extent this is already happening. For example, the DART asteroid-deflection test mission will be using the NEXT-C ion engines developed for the canceled ARM mission.

Exploration is what NASA does. But "space is big," to quote the *Hitch-hiker's Guide to the Galaxy*.[6] The Solar System contains nearly 200 worlds—celestial bodies made round by their own gravity—and millions of smaller bodies. All the planets have been visited at least once (whether or not you count Pluto as a planet). Yet every visit shows how little we understood before. Only in the last decade or so did we realize that several moons of the giant planets are "water worlds" with liquid oceans beneath tens of kilometers of ice. They may even be life bearing. Similarly, that a distant cold world like Pluto could have a complex active geology and atmosphere is also a new and largely unanticipated discovery. There is plenty of room for more exploration. From news reports it may seem that we are exploring the Solar System at a pretty impressive pace, but our actual rate of exploration is pretty slow. Worldwide there are only about 20 probes sent out each decade. At that rate it will take a century to send even one spacecraft to all the Solar System worlds. We need to scale the exploration rate way up if we want to know what the resources of the Solar System truly are within a lifetime. Governments can leverage NewSpace and miners to lower the cost of these expeditions so that we can explore many strange new worlds quickly.

In the same 1962 speech at Rice University in Texas in which President John F. Kennedy called for the United States "to put a man on the Moon . . . before this decade is out," he also said: "There is no strife, no prejudice, no national conflict in outer space as yet."[7] Remarkably, President Kennedy's pacific description of space has held true for over 50 years. His vision is about to be tested as never before. Policy, regulation, and diplomacy will soon come to be central to space. The resources of space are truly vast, but they are concentrated in a small number of locations; at first, only a tiny fraction will be accessible. Policy makers have not paid serious attention to these resources because they seemed far distant, more like tales of El Dorado than real concerns. Highly concentrated resources mean that there will be competition for who gets to use these resources and for what. As Karen Cramer, a space lawyer, has said, "Mining, astronomy, geology, solar power, manufacturing, and landing rights are not all compatible."[8]

The number of players in space, both countries and companies, is growing rapidly. At the moment the stakes are objectively small but, like the inflated share prices of many an IPO, they are valued for what they may become, not what they now are. Who will arbitrate disputes among them? There is now no institution with even nonbinding power to arbitrate over these rapidly approaching disagreements, and the only accepted laws are the few principles laid down in the 1967 Outer Space Treaty. There could be tense times ahead.

Policy issues will soon need to be tackled. Working out how to resolve these disputes requires specialists in policy. They have experience turning technical details into a set of issues that lawmakers can use. I had no exposure to policy making until I gave a lunchtime talk to a group at Harvard on why I thought there would be problems. This lunch discussion was organized by Alanna Krolikowski, a policy expert on Chinese aerospace and currently a professor at the Missouri University of Science and Technology in Rolla. Space activities can be grouped into four areas. Mnemonically we called them sales, security, science, and settlement.[9] By sales we mean space commerce—greed, in our term here. Security is finding and removing the killer asteroids—that's fear. Science is what we already do—what we call love in this book. Settlement is the idea of humans becoming full-time residents of the whole Solar System. As all of the four will use the same technologies, they help one another in that sense. As most people thinking about space resources are technologists, the very idea that the goals may conflict with one another hadn't really been raised. But reality may not be so rosy.

Alanna, being a policy expert, identified the competing goals among our four areas. Sales will want to extract as much wealth as possible as quickly as possible and will chafe against any restrictions. Security cares only about being able to find and deflect or destroy killer asteroids; all the rest is discarded. Science mostly wants to preserve pristine conditions for study, except when it wants to build enormous telescopes using space resources. Settlers want to use all the resources they can get, but should also want to conserve them so that they have a long-term future. Who arbitrates when these goals conflict? That's where policy experts can help in drawing up realistic options.

To address issues arising from asteroid and lunar mining, the Colorado School of Mines has held an annual Space Resources Roundtable since 1999, in collaboration with other organizations.[10] There are few rules about these resources as yet. Many budding asteroid miners want some sort of regulation, but there is no regime of mineral rights, however temporary, that would encourage investors to back a mining operation.

Regulation is an idea that many space entrepreneurs, like their Earth-bound counterparts, react against instinctively. Yet some regulation can be necessary for a market to function. There are regulatory issues that need to be handled soon for which we don't yet have solutions. Safety is one. Once there are tourists and researchers in space, not to mention repair crews to tend to expensive mining equipment, there will have to be safety regulations as well as provisions for emergency services. Public safety is also an issue. If an enterprise moves an asteroid toward Earth for more convenient processing, what rules must it follow to avoid an unintended impact with our planet? ("Environmental impact" indeed!)

Then there are cultural values. Strip-mining the Moon for helium-3 would produce large scars on the Moon that would be clearly visible from Earth. Is that okay? Or a lunar base might have bright lights shining around it at night that will be visible from Earth. Is that inspiring? Or is it sacrilege? Large constellations of thousands of satellites will soon provide broadband worldwide, to the great benefit of billions. At the same time these satellites may be quite visible to the naked eye at dark sites; they will certainly impact astronomy from mountaintops.[11] Does the benefit outweigh the harm?

Even art could be a problem. The now very successful start-up company Rocket Lab launched a shiny "disco ball" into orbit on its first satellite launch on 21 January 2018. The company described this as a sculpture it named *Humanity Star*. The meter-diameter ball would be the brightest thing in the night sky. The goal is to "make people look up and realize they are on a rock in a giant universe," CEO Peter Beck said.[12] That may sound beautiful to you—art is subjective, after all. But astronomers were the first to object: "This is stupid, vandalizes the night sky and corrupts our view of the cosmos," tweeted Columbia University astronomer David Kipping.[13] But who has the power to stop them? *Humanity Star* was in a low orbit and reentered after only a few months. The idea of "astral art," though, will be hard to stop now that it is feasible. There is no legal means of stopping it. It also sets a precedent for advertising.

Real "astral advertising," or "astrotising," is coming. The idea of space advertising goes back at least as far as Robert Heinlein's 1950 book *The Man Who Sold the Moon*. This is no longer fiction. With cheaper access to space it is becoming a reality. Will it be "Ad Men Ad Astra"? Our view of the sky may become obstructed by the advertising logos of corporations.

We might see advertising spread across the Moon. Would we like to see the Nike swoosh on the Moon? Every night? Forever? Who is to stop them? The Moon has spiritual values in many cultures. In Japan *Tsukimi-dai,* or Moon-viewing platforms, have been created for centuries. Margaret Race of the SETI Institute notes that many Native American cultures value the Moon highly.

Vladilen Sitnikov, CEO of the Russian company StartRocket, says, "It's human nature to advertise everything. . . . Brands [are] a beautiful part of humankind," and anyway "there are currently no laws against their ambitions."[14] Libertarian scholars J. H. Huebert and Walter Block argue that there should be no limits on space advertising for free-speech reasons. Not everyone agrees. In the United States there already is an explicit law against "intrusive" space-based advertising. That was all that was needed when the United States was the only capitalist space power, but now there are many ways to get to space. Who has the power to stop space advertising? As of now, no one.

We don't have to despair. Similar situations have been faced before and have been resolved well enough. The first steps tend to be voluntary agreements among the various players. The internet is regulated by the ICANN, the Internet Corporation for Assigned Names and Numbers. ICANN is a private-sector body. There is clearly a huge amount at stake, both financially and politically, in governing the internet. Yet the regulation has worked since 1998, and it does so by consensus. Perhaps this is because ICANN was set up to avoid the UN controlling the internet. Maybe this UN "threat" provides the discipline needed. Space may get by with similar structures.

There already are space examples. The question of what to do with communications satellites once they are defunct is handled by the Inter-Agency Space Debris Coordination Committee (IADC). Founded in 1993, it is made up of 13 space agencies from around the globe.[15] One of the IADC's activities is to coordinate the graveyard orbits to which communications satellites are sent when they are almost out of fuel or become obsolete. The IADC is just an informal interagency committee, but it works. Similarly, the International Telecommunication Union (ITU) regulates orbital locations for geostationary communications satellites.[16] It does this by allocating radio frequencies for TV and other transmissions.

Satellites are spaced out around this 24-hour orbit at distances that prevent their signals interfering with one another. Because the orbits where satellites appear stationary above a spot on the Earth must stay within a region only a few kilometers deep and high, this restriction means there are a limited number of slots in which to park satellites. In this way a scarce and valuable resource is managed peaceably. The ITU took up this function in 1993. Although it is now under the auspices of the United Nations, this is a voluntary system. Yet it works well because all the satellite owners recognize that they need it.

Not all space regulation problems have a solution yet. A glaring example is low Earth orbit debris. This debris is a real and current problem, one that could bar everyone's access to space. Technologies to deal with the Kessler syndrome are being developed, but no one knows who will pay to clear up the mess. There is no international body to arrange for mitigation and removal of this hazardous space junk.

As these disputes are inherently international, diplomacy in some form will be needed to create the oversight bodies that we need. The creation of permanent institutions to support and regulate this vast new resource regime should, perhaps, wait until its specific needs emerge and are better understood.

Some form of legal regime may be needed quite early on, however. Long experience on Earth has shown us that whenever there is a rare, valuable resource, there will be disputes. They can be legal disputes or, especially on frontiers, extralegal and often violent. There is no reason to think it will be different in space.

Once a single profitable asteroid-mining venture succeeds, it will start an "asteroid rush." The best ore-bearing asteroids will be rare and suddenly highly valuable properties. All the usual legal, policing, and justice issues that apply in a gold rush will inevitably come into play. We can expect space pirates, rustlers, claim jumpers, and spies quite soon. They won't all be breaking the law.

Space pirates would be those who find out that you are bringing your ore back, conveniently refined already, and who hijack that payload. They could, for example, attach their own rocket to yours and give it an extra push in whatever direction they like. If they operate a long way from Earth no one will see them. Suddenly your cargo ship will stop phoning home. (Or, if the pirates are really clever, they'll send your spacecraft home empty and months later you'll have an unpleasant surprise when you make a trajectory correction that goes haywire because your spacecraft is a whole lot lighter than you thought.) Even if the pirates made only a small change in the velocity of your spaceship, that would soon take it somewhere you could not find it in the vastness of space. All the pirates need to do then is remove any signs that it is your ore and bring it to market as their own. Space piracy is certainly illegal, as you continue to own your spacecraft wherever it is and, it is increasingly recognized, you own the ore you extracted. That doesn't mean it won't happen.

Space rustlers would start earlier. Suppose that, after a lot of prospecting work, you finally find one of the one-in-a-thousand really valuable asteroids. The rustlers don't want to go to all that trouble; they'd rather track where you are going and get there ahead of your mining spaceship. Then they push "your" asteroid into a different orbit, and it is lost to you, but they know its new orbit. Then they mine it and make a lot more money than you would have done because they don't have to pony up the costs of prospecting. Worse, this is perfectly legal because, for now at least, the Outer Space Treaty says that you can't own the original asteroid. Rustlers can also argue that because they moved the asteroid, it is no longer a celestial body and they can own it outright. You have no recourse.

Space claim jumpers would wait for you to start working your chosen asteroid. They'd know this, for instance, by seeing that you've landed a large spacecraft on it, one that's too big for merely prospecting and so must be for ore extraction. Then they can land nearby on the same asteroid and mine it themselves. As you don't have any ownership rights to the asteroid itself, there's nothing legal you can do to stop them. They can get the benefit of your prospecting work for free. All the Outer Space Treaty says is that if they interfered with your mining operations or damaged your equipment, for instance, by digging too close to it, then you might have some recourse.

Space spies could work in one of two ways. If they stole the information that you used to pick out that asteroid, then that is stealing your intellectual property and you have a legal case. Whether that would compensate you for all that you have lost is not clear. Or they could send a small spacecraft to observe your mining operation. They could learn a lot about your techniques and so save themselves the time and expense of developing their own. Even just following your spacecraft saves them the first steps of prospecting. This is perfectly legal. In fact, the Outer Space Treaty explicitly allows inspections. You might, though, be able to protect your novel techniques under patent law. For space activities it clearly won't be enough to file a patent in any one country. The Patent Cooperation Treaty (PCT) of 1970 makes it easy to file simultaneously in all the countries that signed the treaty.[17] Unfortunately, not every country has signed the PCT, leaving open several dozen potential "flags of convenience" to avoid patent protection.

Given these threats to their business, honest miners may start to put markers into their ore to make them identifiable. Traces of some unexpected element or isotope may do the trick. Or, more simply, they can note the precise composition of their ore. Ore will never be pure water or platinum but will contain contaminants. The pattern of these contaminants will likely be a unique fingerprint for each batch of ore being returned (in the case of pirates), or for each whole asteroid (in the case of rustlers). With that kind of branding in place, these nefarious extralegal space activities can be investigated. That means we'll need space sheriffs who can use the best forensic methods to track down who made off with whose valuables.

Asteroids present unique practical challenges to enforcing space law compared to mining sites on the Moon, Mars, or any large body: they are far more remote; they move and can be moved; most are relatively small; and there are huge numbers of them. All these factors make a space sheriff's job much harder, and open up enticing opportunities for those pirates, rustlers, claim jumpers, and spies. Asteroids were surely intended to be included in the Outer Space Treaty. Back in 1967, though, fewer than 2,000 asteroids were known and almost all were large, comparable to the planetary moons then known. Hence no special consideration was given to them. However, now that we know that there are millions of smaller asteroids with potentially valuable resources, we need to take a closer look.

Clearly something must be done about these sharp practices. Otherwise they may strangle the space mining industry at birth. That means bringing in the law. Space conference audiences are, naturally, full of space enthusiasts.

When I start talking to them about how we'll need lawyers in space they groan: "Is there no escape from lawyers?" My reply is that you *really* don't want to live somewhere without lawyers. Just think about the most lawless places on Earth. You wouldn't enjoy it.

It's too late anyhow! The lawyers are already here; space law is a field of legal scholarship. There are several long-standing space law institutions. The International Institute of Space Law (IISL) was founded in 1960, only three years after the first artificial satellite, *Sputnik 1*, was launched.[18] The United Nations Committee on the Peaceful Uses of Outer Space (UN-COPUOS) is one year older still than the IISL, and its Legal Subcommittee is just one year younger.[19] There are even space law textbooks; one of them, by Francis Lyall and Paul B. Larsen, comes in at over 500 pages.[20]

Despite this, space law is underdeveloped, not least because there is almost no case law. Without that it is hard to know what the right laws should be, because there are always complexities that pop up unexpectedly. That is especially true when it comes to the use of space resources. As Scott Ervin, himself a space lawyer, punned, we are dealing with law in a vacuum.[21]

One thing that is clear from all these examples of space skullduggery is that property rights are central to a space mining entrepreneur's concerns. Daniel Faber, a serial space entrepreneur himself, calls it "secure tenure" rather than property rights. Rustlers and claim jumpers can be foiled if secure tenure over entire asteroids is possible. The rights don't have to be forever. They could come with an obligation to invest ("Use it or lose it") and expire after some limited time, perhaps a decade. Pirates become outlaws if the extracted material is clearly owned by the miners. Spies who steal your intellectual property are easier to deal with as they are breaking laws on Earth. The spies who just observe your operations are trickier to deal with under the Outer Space Treaty. Some creative interpretations will be needed if that is to become an illegal activity.

How would a company establish a mining claim under secure tenure? How do you establish secure tenure over a celestial body? The large number of asteroids means that spacecraft visits to any meaningful fraction of them are impractical. If I do the considerable astronomical work to locate an asteroid, pin down its orbit to show it is reachable, find its size, shape, mass, and (surface) composition, but never visit it, is that enough to establish some form of rights to its resources? Or is this just my intellectual property? If I have to visit the asteroid, is it enough just to land on an asteroid and plant a flag? Do I have to do an assay first? Or do I have to return an ore sample to Earth too? Can an entire asteroid be claimed just by returning a very tiny sample? What if the asteroid is hundreds of kilometers across? Are asteroids even "celestial bodies" under the law? After all, if you can move it, is it still celestial?

And, most basic of all: how can claims be enforced? "A right is a remedy" is a commonplace in law school. It means that for a right to have meaning, there must be some sanction on those who violate that right. If you are the victim of space rustlers, is there a penalty against them? Who decides what the penalty is? Under what jurisdiction would the penalty be applied? Who enforces such a ruling? In other words, who is the law in space?

The Outer Space Treaty is currently the only widely adopted general agreement in space law. This is a United Nations treaty agreed upon in 1967. (There is also a Moon treaty, which is much more restrictive on commercial activities, and applies to asteroids as well as the Moon, but this treaty has been ratified only by a handful of countries.)[22] The formal title of the Outer Space Treaty is the "Treaty on Principles Governing the Activities of States in the Exploration and Use of Outer Space, Including the Moon and Other Celestial Bodies." It adopts principles, not laws. Any laws passed by countries that have signed this treaty have to follow these principles. Whether they actually do so or not is a matter of opinion. The Outer Space Treaty has been ratified by 106 countries, including all the presently space-faring countries. But that leaves nearly 90 countries that are not signed up. Flags of convenience could begin to be used to avoid the constraints of the Outer Space Treaty.

Is the mining of space resources creating property, or is it theft? For space miners the three key points in the Outer Space Treaty are these:

- "Outer space shall be free for exploration and use by all States" (article 1).
- "Outer space is not subject to national appropriation by claim of sovereignty, by means of use or occupation, or by any other means" (article 2).
- "The exploration and use of outer space shall be carried out for the benefit and in the interests of all countries and shall be the province of all mankind" (article 1).

"Free . . . for use by all" sounds promising for miners; surely you can keep anything you mine by this wording? But "not subject to . . . appropriation" seems to mean there are no property or mineral rights to be had, at least until you have extracted your ore. That lack of rights is likely to lead to some of the extralegal activities—piracy and so on—I just warned you about. Finally, "space . . . shall be the province of all mankind" is tricky. If only a handful of companies or countries have the ability to mine space resources, how would benefiting "all mankind" actually work? There's no existing international taxation system that could redistribute some of the space mining profits to non-space-faring nations. Even if there were, the tax would have to be low enough not to discourage more entrants by rendering profits anemic; if space mining were overtaxed then it would stop, and there would be no benefit for any of mankind. As usual in politics and the law, there is no simple or final answer.

"Free for use" already has a precedent. The *Apollo* Moon rocks effectively became U.S. government property as soon as the astronauts picked them up and put them in a bag. After all, who could challenge that? The Soviet *Lunokhod* also returned lunar rocks. Those now clearly belong to the Russian government. Three tiny fragments of the Soviet Moon rocks were sold on the open market, a move that clearly treated them as property.[23] NASA unsuccessfully sued a Chicago-area woman who had purchased an *Apollo 11* bag that had some Moon dust in it, implying that Moon dust can be property.[24] This definition seems to be settled law.

Is it the same with asteroids? No one disputes that the dust returned to JAXA from the asteroid Itokawa in 2010 is owned by the Japanese government. Nor will anyone question whether the rocks brought back by *Hayabusa2* and *OSIRIS-REx* belong to the Japanese and U.S. governments, respectively. So, is the issue settled? Because asteroids are much smaller than the Moon, "picking up a rock" on one can alter the asteroid itself. If the asteroid is only the size of a house, then a room-sized boulder can be a big piece of the whole thing. If the asteroid is a rubble pile, you could go on picking up boulders until there is no asteroid left! Then, while you don't own the celestial body, you do own every last piece of it. That's getting tortuous. You could even put the whole asteroid inside a bag and tow it home, the way the Asteroid Redirect Mission once planned to. Is that the same as picking up a rock? If I process an entire asteroid for ore and leave the uneconomic tailings, are they the celestial object, or are they my property? Would destroying a celestial body be a crime? One thing is certain: once you start to think about the legal issues involved with space mining, they get pretty complicated pretty fast.

There could be a sort of sneaky way out of the "nonappropriation" restriction, at least for the smaller asteroids. Space law expert Virgiliu Pop points out that one definition of a "celestial body" is "natural objects in outer space . . . which cannot be artificially moved from their natural orbits."[25] In that case, if we move a body, then we have modified it, and that makes it property. In practice the only objects in space that can be moved are asteroids. Not surprisingly, this idea is not fully accepted in space law. I have a neighbor, a professor at Harvard Law School, who got really excited about this idea. It raises just the sort of questions that lawyers love. Engineers and scientists like myself, on the contrary, tend to bang their heads against walls when such questions are raised. But they have to be thought through if we want a profitable space mining business.

Physicists have good reason to despair at definitions that use words like "cannot be artificially moved." As a physicist, I know that Newton's second law of motion says any slight touch on an asteroid, even throwing a softball at it, produces an (albeit tiny) acceleration, and that changes the orbit. The lawyers will have to decide just how much it has to change its orbit before becoming property. When the Soviet *Luna 2* spacecraft was the first to be deliberately crashed into the Moon in 1959, it changed the Moon's orbit by some miniscule amount. Was the Moon then the property of the Soviet Union from that moment? No one, not even the Soviets, argued that it was. But a small asteroid is much easier to move. If the Asteroid Redirect Mission had succeeded in putting an entire house-sized asteroid into a bag and moving it toward Earth, would that entire small asteroid have then become U.S. property? Very likely it would. The details of how much a celestial body can be moved before it becomes property will be a legal question informed by, but not determined by, physics.

One important idea in the Outer Space Treaty is that of noninterference. The treaty says that states should avoid "potentially harmful interference with activities of other States" and that "maximum precautions may be taken to assure safety and to avoid interference with normal operations in the facility to be visited." What counts as "harmful interference"? No one knows for sure as there has never been a test case. Surely, though, harmful interference would include having a rocket exhaust burn down toward your equipment. Having your equipment bombarded with rocks and dust from a rocket landing nearby could very reasonably be called harmful too. In the low gravity of the Moon, "nearby" could be a few, or even many, kilometers away. On a small asteroid with virtually zero gravity, it could mean anywhere at all on the asteroid.

If so, then these zones would provide a legal way of excluding others from accessing the space resource you are after, at least in principle. Cody Knipfer, at the time a graduate student at George Washington University's Space Policy Institute, and others call these, rather straightforwardly, "non-interference zones."[26] These could be used to take effective possession of the most valuable real estate in space. Beware, though. A land grab approach may not work out so well in the longer term. It sets the precedent that it is okay to take what you want. More valuable assets may then come to light and be seized by others, to your chagrin.

Not everyone agrees. Laura Montgomery, a practicing lawyer who also teaches space law at Catholic University's Columbus School of Law in Washington, DC, says that the Outer Space Treaty wording applies only to states, so commercial companies are exempt, and in any case, it calls only for consultations, not agreement.[27] There's a long way to go before the meaning of noninterference is settled law.

The recent interest in mining asteroids has led to a push to create laws encouraging the commercial extraction of space resources. All these laws are based on the principles in the Outer Space Treaty, or at least they claim to be. The United States was the first to pass such a law, the "Space Resource Exploration and Utilization Act of 2015."[28] Stripping away the necessary (but confusing to those of us who aren't lawyers) wording, it is a very short law. It states that "a United States citizen engaged in commercial recovery of an asteroid resource or a space resource . . . shall be entitled to any asteroid resource or space resource obtained." So the citizen may "possess, own, transport, use, and sell the asteroid resource or space resource obtained." To stay in synch with the Outer Space Treaty, the act also asserts that "the United States does not . . . assert sovereignty, or sovereign or exclusive rights or jurisdiction over, or the ownership of, any celestial body." It does clarify that "space resources" includes water and minerals. It doesn't say what "obtained" means.

Luxembourg followed the U.S. approach in 2017, but with far more detailed legislation.[29] It is careful to allow resources claims only by Luxembourg-based companies with a strong financial base and of good repute, explicitly calling out money-laundering and terrorist-supporting organizations, for example. Amara Graps, a Latvian-American planetary scientist and asteroid miner, says, "It's a great law. This one is more flexible than the US version." In 2019 Belgium signed an agreement to work with Luxembourg on space resources. In 2016 the United Arab Emirates said the country was close to having its own space resource law on the books.[30] As more countries adopt laws, and learn from others, a set of customary space law should emerge. Of course, this was thought to be happening as far back as 1963, so predictions can be way off.[31] The imminent prospect of many players trying to extract these resources should speed things up.

At some point an international body will be needed to adjudicate disputes. What do we have so far? The UN Committee on the Peaceful Uses of Outer Space is the most prominent. It is assisted by an expert unit, the United Nations Office for Outer Space Affairs (UNOOSA). (International bodies do tend to adopt cumbersome names!) UNOOSA already runs a series of Space Law Workshops. There are other organizations devoted to space law with no institutional connection to any governments. The IISL, a nongovernmental organization, describes its key mission as "the promotion of further development of space law and expansion of the rule of law in the exploration and use of outer space for peaceful purposes." It does not take particular positions on space law. Its role is to hold symposia every year, including the Eilene Galloway Symposium, named after a pioneer of space law. IISL is the go-to organization in the field, so it's influential.

A much newer space law group is The Hague International Space Resources Governance Working Group. This group is entirely ad hoc, but its members are well regarded. In the absence of any other framework, they spent three years developing *Building Blocks for the Development of an International Framework on Space Resource Activities*.[32] This compact eight-page document addresses all the topics we've discussed and seems to be setting the tone for the international debate.

Insurance companies will get nervous about asteroid mining. The residue of thousands or even millions of tons of finely ground debris from asteroids has to be contained. Letting this dust disperse into space in the orbit of the near-Earth asteroid is not a good option. As we saw earlier, the solar wind and radiation pressure will spread the dust out along the orbit, producing a fabulous meteor shower. But this would also create a new hazard for Earth-orbiting satellites. That raises a liability question. Suppose my delicate half-billion-dollar communications satellite has a pebble of asteroid debris shot through it at 36,000 kilometers per hour, destroying its electronics. Is that just bad luck, or is the mining company at fault? Some NASA satellites already take special precautions during the most intense meteor shower, the Leonids. The *Chandra X-ray Observatory*, which I worked with for many years, turns its backside to the Leonids each year to protect its delicate X-ray mirror at the front.[33] A geostationary communications satellite can't do that; its antenna has to stay pointing at the Earth for it to work.

The need to fill the space law vacuum is by no means a far-off concern. A commercial Moon race has begun, with half a dozen companies planning to land there within a few years. Given that the best resources are concentrated in small areas, there is sure to be competition for them.

When the law breaks down, violence is likely to ensue. That could lead to the weaponization of space. If there are disputes on the Moon or on asteroids, and these come to be seen as national security issues, then there will be pressure to up the game. A military organization, such as the U.S. Space Force, might be called on to adopt a policing role. If their targets are pirates, there is likely not to be much objection. So far, so good. But major corporations competing for the same multi-billion-dollar asteroid would raise the stakes worryingly.

Space is already militarized. The National Security Agency (NSA) and U.S. Air Force space budgets combined are (probably) larger than NASA's.[34] They operate fleets of spy satellites gathering intelligence on potential enemies, as do other nations. It tends to be forgotten, but the GPS navigation system that we all depend on is a military program.[35] The most accurate GPS locations were originally encrypted, but this restriction was lifted in 2000. The military value of GPS has led to a lot of duplication. There are independent European (Galileo), Russian (GLONASS), and Chinese (BeiDou) navigation systems. But militarization is not the same as weaponization. While nuclear weapons are excluded from space by the Outer Space Treaty, weapons in general are not.

War is illegal. This radical idea was first proposed by the (more famous than read) philosopher Immanuel Kant in his 1795 essay *Perpetual Peace*.[36] The principle was actually enshrined into international law by the much-derided Kellogg-Briand Pact (sometimes known as the Pact of Paris) in 1928. Law professors Oona Hathaway and Scott Shapiro have revived discussion of the legality of war.[37] However idealistic that (sadly) still sounds, war in space really is illegal under the Outer Space Treaty, at least on a solid body in space. As the treaty clearly states: "The establishment of military bases, installations and fortifications, the testing of any type of weapons and the conduct of military maneuvers on celestial bodies shall be forbidden" (article 4).

Asteroid mining may make the prospect of war worse. Whenever I mention mining asteroids for platinum group metals to people, their first reaction is to ask, "What about rare earths?" The rare earth elements are vital to modern electronics. That effectively means that a modern economy would collapse without access to them. Rare earths command only modest prices, around $50 per kilogram.[38] That's about 1,000 times cheaper than the platinum group metals. That makes them unlikely to be commercially viable products to obtain from space. And, despite their name, rare earths are not particularly rare, just hard to mine. Rare earths are particularly environmentally destructive to produce, which is why U.S. production was shut down.

Strategically, though, the near monopoly on rare earth production by China has caused worry. The restrictions that the Chinese imposed on exports in 2010 showed how sensitive the market is to disruption in supply. Prices trebled for a short time. In practice China's export restrictions were ineffective and were dropped in 2015.[39] The strategic lesson may be important though. U.S. and other countries' interests would be well served by ensuring an independent supply of these strategic elements. There is a region on the Moon called the KREEP Terrane that has unusually high concentrations of rare earth elements.[40] (That's what the REE stands for; K and P are the chemical symbols for potassium and phosphorus.) Could the Moon and asteroids provide a benign and environmentally friendly source of these strategic elements at a higher, but still reasonable, price and so retire worries over their supply? Or will rare earth ore be so hard to find that it encourages conflict? It would be smart to keep these issues in mind as space resources for more obviously profitable materials begin to be developed.

If asteroid resources become a matter of importance to nations on Earth, then geopolitics (if *geo* is still the right prefix) will enter the picture, and disputes and tensions could get much worse. If the asteroids become the primary source of some strategically important resource, then they may be seen by some players as worth fighting for. Space saboteurs could be used to prevent you from mining your chosen asteroid by making it radioactive or by blowing it up into a shower of small fragments. Sabotage could then be a casus belli. If there is enough at stake, then that war may not stay in space.

Some weapons are clearly excluded from space by the Outer Space Treaty: "States Parties to the Treaty undertake not to place in orbit around the earth any objects carrying nuclear weapons or any other kinds of weapons of mass destruction, install such weapons on celestial bodies, or station such weapons in outer space in any other manner" (article 4). So exploding a nuclear bomb to make an asteroid radioactive is clearly not allowed. Blowing the asteroid to pieces, even with a simple kinetic impactor (a "hammer") may also count as using a weapon of mass destruction, but that's less obvious. Like the ban on appropriating celestial bodies, the ban on weapons of mass destruction turns out to be less clear-cut than you would think. The question is "What is a weapon?" When it comes to stopping a killer asteroid from hitting the Earth, a nuclear bomb may be the only practical option. Should the Outer Space Treaty be reinterpreted to state that such Earth-defense nukes are not really weapons but tools? They could be either, depending on the purpose for which they are used.

Many space technologies are "dual use" like this. Another example is the desirable goal of having the capability to de-orbit or refuel satellites to reduce space junk or to extend their useful lives. The same technology can also be used to inspect spacecraft to learn their capabilities or to immobilize them to prevent their use in a war. As William Shelton, former head of the U.S. Air Force's Space Command, puts it, the difference between a servicing spacecraft and a weapon is merely "a change of intent."[41] A more dramatic example would be turning a space-based solar power station into a weapon against places on Earth simply by adjusting its focus.

The United States created a "space force" as the sixth branch of the military in 2019. Are its members going to be the space sheriffs of the future? No, at least not right away. At this point the U.S. Space Force is more of a bureaucratic reorganization, bringing the satellites of the U.S. Army, Navy, and Air Force under a single command. President Trump stated that the goal is "American dominance in space."[42] That aggressive rhetoric suggests that eventually the space force may add offensive capabilities to defend against attacks on U.S. satellites, including the GPS system. Those capabilities may then lead to the ability to interfere with or destroy the satellites of other nations. The use of the phrase "American dominance" is not going to be welcomed by other nations, space-faring ones or not. Stephen Kinzer of Brown University says that goal "will provoke ruinously expensive competition among nations."[43] Even the name "space force" conveys that message. *Apollo 11* astronaut Buzz Aldrin said, "I have thought for some while that a better name would be Space Guard, because it is more deterrent."[44]

Ever since the early seventeenth century and the writings of Hugo Grotius, much effort has gone into trying to decide what makes a war just or unjust.[45] This continues in the context of space. A team from the McGill Centre for Research in Air and Space Law started the *Manual on International Law Applicable to Military Uses of Outer Space*, or *MILAMOS*, project.[46] *MILAMOS* aims to answer the question "When and under what circumstances it is lawful for nations to resort to hostilities in or through space?" Perhaps inevitably, there is a rival, the *Woomera Manual on the International Law of Military Space Operations*. Both groups feel that war in space is inevitable. "Conflict in outer space is not a case of 'if' but 'when.' However, the legal regime that governs the use of force and actual armed conflict in outer space is currently very unclear, which is why the Woomera Manual is needed," says Woomera founder Professor Melissa de Zwart, dean of the Adelaide Law School at the University of Adelaide.[47] The annual Space Security Index (SSI), which includes researchers from both McGill and Adelaide, tracks developments in the field.[48] In 2018 researchers noted that several countries now have the ability to shoot down satellites in low Earth orbit. Will international regulation be sufficient to reduce the risks of conflict in space or over space assets?

Despite having just talked about war, most of this book is full of sunny optimism. Hopefully, though, I've put some realism into the too often

Pollyannaish pronouncements you will find about space trillionaires, while also giving you reasons to lend that optimism some credence. Still, failure is an option. Relying on greed to expand into space means that the possibility of failure is ever present. Capitalism is merciless to those who fail to make a profit soon enough, large enough, and consistently enough. To a scientist like myself, not having solid proof that a venture is viable is a real turn-off. It took me a while to realize that to entrepreneurs a venture only has to be plausible. If they wait until a profit is an ironclad certainty, then they'll be too late. Someone else will have taken the risk already. Carissa Christensen, founder and CEO of Bryce Space, says that three-quarters of all venture capital–financed companies fail.[49] As Erika Ilves points out, visionary entrepreneurs need nerves of steel.[50]

We have already seen several examples of failure or near failure in the NewSpace movement. Ilves's own start-up, Shackleton Energy, did not attract the major investment needed to mine the Moon, but she went directly on to begin another start-up, DeepGreen Metals, to mine the deep ocean, while her co-founder, Jim Keravala, started OffWorld to concentrate on mining automation. The first two asteroid-mining companies, Chris Lewicki's Planetary Resources and Rick Tumlinson's Deep Space Industries, were both bought out within two months of each other at the end of 2018. Other companies will step up to take the risk; and maybe this time they'll win. In fact, there are already several other companies planning to get asteroid mining going. Members of the Planetary Resources team quickly founded a new venture, First Mode, while Daniel Faber, formerly at Deep Space Industries, founded a new company, Orbit Fab, Inc., which launched hardware to the ISS within months. In capitalism, failure is not a bug, it's a feature. All these examples show that the losers don't have to lose forever. For the entrepreneur "success is not final, failure is not fatal: it is the courage to continue that counts" as the well-known quote goes. (This epigram is often wrongly attributed to Winston Churchill, surely because it does, after all, sound like him. Most likely, though, it comes from a 1938 Budweiser beer ad![51] That sounds appropriate for entrepreneurs.)

State-funded projects also fail. But when they do there is no competitor around to keep things moving. When the *Challenger* exploded 73 seconds after takeoff in 1986, it was 32 months before a shuttle flew again. Yet the problem that caused the explosion was clear very soon—the O-rings connecting the segments in the stack making up the solid rockets are supposed to be flexible to ensure a tight seal. But in the subfreezing temperatures at the *Challenger* launch, they became brittle. This was made obvious in the dramatic demonstration by Nobel Prize–winning physicist Richard Feynman, who was on the Rogers Commission that investigated the disaster. During one of the commission's public sessions he dunked some of the O-ring material into his glass of iced water and then snapped it like a ginger cookie.[52] The problem had been well known to the engineers who worked on the solid rockets. They tried to warn NASA officials not to launch that day, but to no avail.

Contrast this with what happened when the one of the three rockets on the first *Falcon Heavy* flight failed to land on the drone ship *Of Course I Still Love You*. The same day Elon Musk tweeted that it ran out of fuel a few seconds early. "The fix is pretty obvious," he ended.[53] SpaceX simply made the fuel tank larger and tried again a few months later. This time it succeeded. The obvious difference compared with the *Challenger* is that no people were on board the *Falcon Heavy*, so there was less risk to life. For SpaceX there was still significant risk though. SpaceX lost millions of dollars of equipment on each failure, and its reputation as a reliable launch company was at stake.

Reality, even if things go well, is never without problems. In planning any large enterprise there's a well-known tendency to project a completion date that turns out to be overly optimistic. Anyone who has dealt with contractors renovating a kitchen knows how things start off well with a rapid demolition job on the old kitchen, and then somehow the project inevitably stretches out horribly. If you lay out a timeline of when you expect things to be done, then put in the real dates of when they actually get done, you see that everything, always, "slips to the right."

Space projects are infamous for being extreme examples of this "slipping to the right." A startlingly egregious recent case is the NASA flagship mission, the *James Webb Space Telescope*, or *JWST* for short. When it was originally conceived in 1997 it was projected to cost only half a billion dollars—very cheap by NASA standards—and to fly in 2007. To date it has cost nearly $9 billion and it won't launch until at least 2021. This is a minimum of a 14-year delay. The whole telescope was designed only for a 5-year lifetime, so the delays are almost three times longer than its planned operations. (Although 10 years and more of operations may be possible with careful husbanding of *JWST*'s onboard resources.) Virgin Galactic is about a decade behind where it first expected to be.[54] The SpaceX *Falcon Heavy* first flew in 2018, 5 years after its initial target date.[55]

Similar slips will happen in my optimistic estimate of when asteroid mining will begin. The infamous "unknown unknowns" will crop up. On the plus side, greed will help keep people on schedule.

Another looming danger, and the first question I always get when I talk about bringing precious metals to Earth from asteroids, is "Won't that crash the market?" Will precious metals cease to be precious? It's a good question. Certainly, small changes in supply can make a big difference to prices. In 1979 just a 4 percent drop in oil production, due to the Iranian revolution, led to a doubling of prices. Platinum prices already fluctuate strongly.[56] They are particularly sensitive to economic recessions. In the 2008 recession the price dropped within six months from almost $80,000 per kilogram to $30,000. That kind of change makes it hard to plan years ahead, as any asteroid miner would have to do. If we bring back too much, too quickly, we could end up depressing the market and turn our profit into a loss. That would not be good. Is there any way to estimate the effect of adding asteroid precious metals to the market?

Economics is not my strong point. After all, I didn't go into astrophysics for the money. I knew I could use some help here. So I went across the Charles River to talk to economists at the Harvard Business School. It took me a while to find a sympathetic ear; space mining was still a really fringe topic when I first tried. It didn't seem sufficiently businesslike to them, I guess. Eventually I found help from Matt Weinzierl. Matt is a trained economist, so he understands the problem, but he specializes in tax policy, so this wasn't really his thing. Still, he has been writing case studies on NewSpace companies, so he was interested. He consulted some colleagues, in particular Martin Stürmer of the Federal Reserve Bank of Dallas. I was surprised when he told me that economists have only just started to look at the question.

Matt points out that we can think about the problem if we know how "elastic" the demand for the material is—that is, how much it responds to the increase in supply. For example, if a large increase in the supply of platinum inspires people to find an array of new uses for platinum or to dramatically expand the scale of existing applications, then the price need not fall very much, even if we bring back a great deal from space. That's the case of very elastic demand. But if demand is *inelastic,* then there's a real possibility that a large increase in supply will crash the market. The problem is that there's a great deal of uncertainty over how elastic demand for a given resource will be in the future. And that uncertainty deters investment. Carol Dahl of Luleå University of Technology, Sweden, and the Colorado School of Mines has taken a more careful look at the problem with two colleagues, Ben Gilbert and Ian Lange.[57] They find that if there are only moderate and gradual increases in the supply of metals from asteroids, we would not expect a large effect on prices even if demand is not very elastic. This suggests that mining companies may be willing to invest, especially if they can control the supply from space to a certain extent.

About 200 tons of platinum are presently produced annually.[58] At a price of $50 million per ton, that is a market of roughly $1 billion per year. If asteroids supplied just 20 tons a year, the amount we might find in a 100-meter-sized metallic asteroid, then that would be only a 10 percent increase in supply. That's the sort of increase that Dahl and colleagues find to have only a small effect on the price. In that case, we could rake in virtually all of the $100 million a year that we hoped for. We might get a similar cash flow from palladium, which is about as expensive as platinum, and very useful. Alas, even $200 million a year in revenue does not make for space trillionaires. Unless asteroid mining can get up to billions per year in revenue, there's not going to be a giant space mining economy.

Eventually, I am confident, asteroid mining will start to pay off, and once it does, human wealth will grow to levels we will find hard to understand. There may then be space trillionaires after all. We should think about what kind of future we want as we craft laws and build the governance structures that will guide our expansion into the Solar System. Now is a good time to do that, while we know something about the possibilities, and before there are strong vested interests.

10

In the Long Run

Will humanity be restricted to Earth forever, or will we be able to spread through the Solar System and—perhaps—beyond? In other words, do humans have a future in space? If we do, the vast resources we can make use of thanks to asteroid mining will have to be central to that future. To survive outside of Earth humans must be able to live in space and so must find the resources out there to sustain us. Moreover, those resources must be harnessed at a cost that allows their continued and increasing use. We express this need today by saying that mining them must be profitable. Asteroids are by far the largest reservoir of useful and accessible materials in space. As we've seen, these include water—for living, radiation protection, and rocket fuel—construction materials—rock and iron—and the specialized valuable materials—rare earth elements, platinum group metals—that sustain our advanced technologies. If we can mine asteroid materials profitably, then we can become a Solar System–spanning civilization.

Harry Shipman, an astronomer, stripped the problem down to two essential questions in his 1989 book *Humans in Space*: Can we support life in space without supplies from Earth? Can we make a profit doing things in space, so that our presence in space can grow without limit? We don't know the answer to either question yet, but we can look at how things would turn out depending on those answers. Shipman's options fit conveniently into the old fallback of management consultants, the two-by-two matrix.[1] That gives us four possible futures in space. Table 1 lays them out. Today space is not profitable, nor can we live off space resources. Space

Table 1. Our four possible futures in space

	Can't live off the land	Can live off the land
Not economically profitable	Today's space program *Antarctica*	*Kalahari*
Economically profitable	*Offshore oil platforms* *Yukon*	Growing space settlement *California (or Bolivia?)*

thus resembles the case of Antarctica. Having just one factor but not the other does not help. People live on profitable offshore oilrigs, but only temporarily and not off local resources. People also live in the Kalahari Desert full-time, but very little profit is made there and the Kalahari economy is not booming.

This book has been about the fourth case, in which we can make a profit *and* we can live off the land. The space economy could then grow exponentially, as economies do. The model you'll hear talked about most by space expansion enthusiasts is how gold rush California grew into the economic powerhouse of today. This is certainly a great example. It won't necessarily work out that way in space, though. Bolivia is a counterexample. The Potosi silver mines in Bolivia produced enormous profits, and it is certainly possible to live off the land in Bolivia. Yet Bolivia did not have the spectacular economic growth of California. Why not? The Potosi mines were run by the Spanish state and staffed with conscripts from the surrounding regions.[2] These seem to me to be unlikely conditions to produce a dynamic Bolivian economy. While space settlements need both profitability and the ability to live off the land, they will depend on other factors too.

The way asteroid resources are distributed between a modest number of near-Earth asteroids and enormous numbers of Main Belt asteroids implies that there will likely be three ages of our future history: scarcity, abundance, and, potentially, exhaustion. The era of scarcity will play out in the next few decades; the era of abundance over the next few centuries; and, if the second period comes to pass, an era of exhaustion in the next millennium. Each of these three ages will bring up moral questions that we will have to face.

On a timescale of decades we will be in an era of scarcity for space resources. Only a modest number of near-Earth asteroids and the Moon can
provide profitable space resources for now. Inevitably, high value and
scarcity will lead to disputes. How we deal with this stage could set the
tone for many decades down the line.[3] Scarcity raises the probability of
piracy and other extralegal activities, as well as war. These are very real
potentials that most of us would want to avoid. The distribution of the
gains will need to be seen as just, or else the corporations exploiting the
resources will face troubles back on Earth. A just regime is a stable regime.

So long as profits can be made at this stage, more capital will be invested
and technology improved, letting us mine many more of the near-Earth
asteroids. That will ease the scarcity, somewhat, and will start to make the
in-space economy significant, providing the material basis for permanent
habitats in space.

On a timescale of several centuries, we can look to an era of abundance.
That's because, if profits are made, then eventually the many asteroids of
the Main Belt will be mined. At that point the resources available to humanity will become truly vast, millions of times what we can access on the
surface of the Earth. If the human population does not grow so fast, then
we could each be a thousand times richer than we are today for centuries.
Robots are likely to do most of the work. There might be something like
a universal basic income for people, but set at a level we would consider
"rich" today. This would be the "post-scarcity" dream from *Star Trek*
come true. No one would want for material things, if we organize ourselves
appropriately.

What kind of economy will we have in space a century or two from now?
I like to think about what the grizzled California '49ers of the 1849 gold
rush would have made of the California economy 100 years or so later, in
1959. That later economy was based on agricultural exports, entertainment,
and aerospace. Disneyland had just opened. The most far-thinking '49ers
would have understood that California could export food, but only if there
were a railroad crossing the entire North American continent. The first
such line was a decade in the future, and many thought it impractical. As
for exporting entertainment, what could that possibly mean? Hollywood
was unimaginable. The Lumière brothers were 30 years from showing the
first-ever movie. And "aerospace" would get at best a baffled look from a

'49er. They knew about hot-air balloons, so flying in the air made some sense, but the first heavier-than-air flight, by the Wright brothers, was 50 years away. And what would they make of the "space" part? Surely, they would shake their heads sadly at our craziness. True, Jules Verne was to publish his book *From the Earth to the Moon* only 16 years later, in 1865. But that was a fanciful tale that no practical person would take seriously. In fact, the first satellite to orbit the Earth came only in 1957, more than a century after the '49ers. My guess is that we are as unlikely to be able to imagine the economy of the twenty-second century as those '49ers trying to imagine that of the twentieth.

People would still work in this potential post-scarcity future, if we can bring it about. They would not work to keep themselves alive or to pay the mortgage. So what would motivate them? Greed has been done away with; up to some pretty huge amount you can have whatever you want. Fear has been minimized, though it can't be eliminated entirely. Love remains. People would do what they love. Their "work" might be playing in sports leagues, for example. The lives of the English gentry in the eighteenth century may be a guide. These are the people Jane Austen wrote about. A small percentage of the English population did not work, or perhaps performed very light duties in the church. Most were happy to potter about, while some recorded the comings and goings of plants and birds in their gardens, or thought about probability and the proof of God's existence, thereby incidentally founding new branches of science.[4] Similar people of leisure in Chekhov plays went in for ennui or shooting people. Let's hope the English model prevails.

In this world many people would also work for reputation, on a quest for renown and honor among their peers. Honor and renown may sound like irrelevant knightly virtues from medieval tales. Yet they are not so strange. They are the motivations of scientists today. Most scientists could get another job outside of research, most likely one where they would make quite a lot more money. But we don't. That's because we make enough to be comfortable, but not rich, and we value these other rewards and our love of the quest for knowledge more highly.

This view may be too Panglossian.[5] Most of the wealth could end up being gathered into the hands of a few space trillionaires. They could amass power that exceeds anything an Earth-bound government could muster.

And the remoteness of the asteroids will make it hard to enforce laws designed to control them. At some point the unfettered capitalism that harnessed this plenty, greed, will cease to be the right tool for promoting our well-being. Whether we will choose post-scarcity or space trillionaires is a big political challenge.

As in many *Star Trek* episodes, ethical dilemmas will still have to be dealt with even in this post-scarcity future. The Martian moons, Phobos and Deimos, will be accessible to us. In fact, they may make a good base for mining the Main Belt. In that case going to the Martian surface will be a relatively small step. Permanent settlement there will surely follow. That raises the idea of environmental protection in space. Is any protection due to these (presumably) dead worlds? If so, what places should we set aside as wilderness, or as the common heritage of humankind? The largest asteroid, Ceres, perhaps? The largest mountain in the Solar System, Olympus Mons, or its largest valley, the Valles Marineris, both on Mars? Should these places be preserved in near-pristine condition? In the more distant future, would it be moral to strip-mine Saturn's rings for their water until they are all gone? (Doing just this is part of the economy in *The Expanse* science fiction books and TV series.) Would that be a crime? If we do find life on Mars, should we preserve it? Would destroying an entire tree of life, even a small one, be a greater tragedy or crime than causing the extinction of one species? Carl Sagan once wrote, "If there is life on Mars, I believe we should do nothing with Mars. Mars then belongs to the Martians, even if the Martians are only microbes."[6] Obviously not everyone agrees.[7]

On a timescale of centuries, Mars may look like small potatoes. Jeff Bezos talks about a trillion of us eventually living in space.[8] That's a bit over 100 times today's population. Is that a good idea? With 10 million times more resources in space, it means a trillion of us could each use 100 times more material resources for 1,000 years. That seems a lot, but we would need those resources. The planetary area of the Moon and Mars together are about the same as the land surface of the Earth; if we could make Venus habitable too, that would add another 2 1/3 times as much as Earth. So we could house just over 3 times today's population on these celestial bodies at the same density as on Earth today. But the remaining 960 billion people will have to live in giant space habitats, very much in the Gerard O'Neill style. In those habitats, instead of having our land, our gravity, and our air for free, they must all be engineered. And they must be engineered to last, like medieval cathedrals. How our machines age in space on a centuries' timescale is unknown; the Space Age isn't that old. The NASA *Voyager 1* spacecraft is still operating after more than 40 years, but most of our satellites fail much faster. That won't be an option for a spacecraft with a million passengers. To house a trillion people may require the exploitation of all of the resources of the Main Belt to achieve Bezos's goal. Is that the best way to spend those resources?

Will Earth itself then be transformed into a park, a playground, a garden of Eden? This is Bezos's dream and it sounds delightful. Would it ever really make sense to make everything in space though? Will we import dishwashers from space? That's hard to believe right now; but we do import T-shirts from across the globe, something that would have seemed hopelessly impractical not so long ago. We should, perhaps, reserve our judgment on that until we see how the space economy takes shape.

Suppose we do export industry to space and make Earth into a paradise. People will really want to live here. That will lead to a lot of tricky questions. How large a population could Earth support if it is to be a garden? Ten billion, like today? A few hundred million, as in pre-industrial times? Or even more billions than today? How will they be chosen and by whom? Will they be an elite? If you can pass on your Earthly estate to your children, as seems so reasonable today, then won't it follow that Earth-dwelling will become hereditary? Will there be rich people, even then, who alone have the privilege of living on Earth? And then, how is the garden preserved? Will you need a permit to have a child, under threat of being forced to emigrate for doing so without permission? Or, if Earth makes no material goods, then might its inhabitants be held hostage by "spacers" blockading essential imports? The privilege of the "Earthers" then becomes a curse.

As we approach the timescale of a millennium, we will reach the third stage—exhaustion of all obviously accessible space resources. At least we will do so if we let our use of space resources grow without limit, as we have with Earth resources. Millions of times Earth's resources is a vast amount, but it is not infinite. If we keep on doubling our use of iron every 20 years, as we have since the beginning of the Industrial Revolution, then surprisingly soon, within the next 400 years, that path will have brought us to the point of exploiting all of the iron in the Main Belt asteroids. If the space economy doubles at China-like rates, every 7 years, as many space enthusiasts hope, then exhaustion will happen in just 150 years. Your grandchildren will see it coming.

The Solar System is much bigger than the asteroid belt. But going out to the next great store of resources, the Kuiper Belt beyond Neptune's orbit, buys us only a little time, even though it has 10 times the mass of the asteroid Main Belt.[9] (Most of that mass is thought to be ice, but let's ignore that for now.) We gain only three and a bit doubling times and then, after just 60 more years, we are stuck. Massive imports from the cloud of billions of comets surrounding our Solar System are not practicable with today's science. The journey times become decades to centuries. The situation is even harder for going to other star systems. Even with fusion rockets the journey times to them would still be centuries, and the maximum payload we could return would be tiny in relation to an economy a million times larger than that of the Earth's today. Economics Nobel laureate Paul Krugman showed long ago that even setting prices for interstellar imports gets very tricky when the slowing of time due to Einstein's relativity for travelers near the speed of light becomes important.[10] We do have natural imports from other stars in the shape of interstellar asteroids like 'Oumuamua, but only as much as the world used in the year 1800 and far too little to feed such a vast economy. And what then? Recycling will help delay this crisis. Recycling people's space habitats, though, will run into some serious resistance, I imagine. Some resources, once used, are truly gone. Whatever we use for rocket propellant is irrecoverable. Unless we come up with a new way of doing business, we will have a giant economic crisis.

Surely 400 years of new science will save us from the crisis of exhaustion of the Solar System's resources? After all, 400 years in the past, Galileo had only just turned his telescope to the skies. He discovered wonders, but they pale in comparison to what we have found since. Won't it be the same 400 years in the future? Unimaginable scientific breakthroughs are bound to give us a "Get out of jail free" card, no?

We may not be so fortunate. Back in 1965, during a rush of discoveries in high-energy physics (and long before his demonstration of the shuttle's O-ring problem), the Nobel Prize–winning physicist Richard Feynman said: "We are very lucky to be living in an age in which we are still making discoveries. It is like the discovery of America—you only discover it once. The age in which we live is the age in which we are discovering the fundamental laws of nature, and that day will never come again. It is very exciting, it is marvelous, but this excitement will have to go."[11] If he was right, then all the discoveries there are to be had may well have been made long before 400 years from now. And they may not include any faster-than-light propulsion that would let us reach the stars. I find that a saddening thought, but it is possible.

Feynman could be wrong though. What might happen? Our best hope to escape the Solar System will be finding new physics, like the warp drive of *Star Trek*. Is there hope? Any discovery that will help is likely to come from high-energy physics and cosmology. Here are some far-out ideas that illustrate the kinds of surprises that just might be in store.

Out beyond Neptune, in the Kuiper Belt, a "Planet 9" may lurk. Mike Brown of Caltech, self-described "Pluto killer," proposed that such a Planet 9 could explain the peculiar orbits of the large Pluto-like worlds out there.[12] (It is the many "plutoids" that led to the demotion of Pluto to dwarf planet status.) Planet 9 is usually assumed to be a stray planet that was ejected during the turbulent early life of some other solar system and wandered the Galaxy for eons before being captured by the gravity of our Solar System. But there is a stranger possibility: Planet 9 could be a mini black hole created during the Big Bang, as some have proposed.[13] Although it would be some 10 times as massive as the Earth, it would be only about the size of a baseball. If Planet 9 does turn out to be a black hole, we could go out to it and perform experiments on a black hole for the first time. That may sound scary, but at least the outer Solar System is the safest place we could possibly make such experiments. Black holes are thought to be the simplest things. A mass, a spin, and an electric charge are the only properties they possess. Given a chance to experiment on them directly, perhaps we will find that black holes have more complex properties than we can measure from afar or that our present theories predict. Black holes are extreme distortions of space-time. Experimenting on a black hole may just help us get our own warp drive.

In principle it is possible to warp space-time to let your spaceship arrive somewhere faster than would have been possible at the speed of light. You don't actually go faster than light; it's space-time itself that changes. The idea of this "warp drive" came from theoretical physicist Miguel Alcubierre of the National Autonomous University of Mexico (UNAM) in Mexico City, in 1994.[14] Unfortunately, his drive would require some new form of matter with negative mass, which isn't known to exist. And yet . . . it has been suggested that the dark energy that is accelerating the expansion of the universe could be explained with negative mass particles.[15] To be fair, this is one among hundreds of such suggestions and comes with its own problems. Another tricky problem with the Alcubierre drive is that space-time is also very stiff and hard to bend. Doing so would take enormous amounts of power. One source of enormous power could be a "quark nugget."

Quark nuggets are another completely hypothetical form of matter that would also be remnants of the Big Bang. Ed Witten, who works where Einstein used to, at the Institute of Advanced Study in Princeton, New Jersey, suggested back in 1984 that super-high-density "quark matter" could exist in small stable nuggets.[16] He suggested that these quark nuggets might be the mysterious dark matter that troubles cosmologists. That dark matter has five times more mass than the normal matter of which stars, planets, and you are made. Yet all we see of it are its gravitational effects. Quark nuggets are just one of hundreds of suggestions of what the dark matter may be.

If they exist, then quark nuggets would have the extremely high density of neutron stars. Neutron stars are as massive as the Sun, but just a few kilometers in diameter, the size of a modest city. A small dice-sized chunk of neutron star matter would weigh in at half a billion tons.[17] Gravity on the surface of a quark nugget would be 160 million times stronger than we experience on the Earth's surface! In that strong gravity, dropping just 1 gram per second onto the surface would release 200 gigawatts, over 10 times the power released by the Three Gorges Dam in China, the world's biggest power plant.[18] That might be enough. One awkward detail is that only neutrons could reach the surface of a quark nugget, but we can always bring along a nuclear reactor to generate those. Since there aren't many free neutrons, quark nuggets are "mostly harmless." If one hit the Earth it would just drop to the center and sit there quietly. Finding a quark nugget to experiment with would surely open up new areas in high-energy physics. You'd have to be really cautious about how you did those experiments, though. Vaporizing your experiment is probably not a good idea!

Some quark nuggets could be out in our Solar System waiting to be found, masquerading as asteroids. If quark nuggets really are the dark matter, then we know how many there must be in our Milky Way Galaxy. Some of those will drift by the Solar System and get captured by straying too close to Jupiter. A preliminary estimate is that there would be about 10 or 100 quark nuggets hiding in orbits around the Solar System with about 1 percent the mass of Ceres, the largest asteroid.[19] An asteroid with 1 percent the mass of Ceres would be about 200 kilometers across. A quark nugget that massive would be only 7 meters across—almost 30,000 times smaller. That is a very small size to detect any further away than the Moon. The chance of one of the 100 possible quark nuggets getting that close is tiny. But as we spread throughout the Solar System, we will surely have vastly better, and more widely distributed, telescopes than we have today. A quark nugget would be perfectly reflective so it would at least stand out because it would have a distinctive color, precisely the same as the Sun.[20]

So it is possible, if only remotely, that a combination of exotic materials out in the depths of our Solar System could give us the keys to escape to another star.

Sometimes it is said that astronomy and particle physics are no longer useful subjects. For example, back in 2002, John Marburger, then the director of the Office of Science and Technology Policy in the George W. Bush White House, said that they "have receded so far from the world of human action that [they] . . . are no longer very relevant to practical affairs."[21] I beg to differ. It could be that, in the long run, astrophysics and high-energy physics are practical subjects, even essential ones.

We began with love, fear, and greed. The last couple of chapters have been all about greed. This is deliberate. My belief is that love's frustration will be requited, and our fear will be assuaged, by greed.

The motto of the *Lunar Reconnaissance Orbiter* camera team is "Scientia facultas Explorationis, Exploratio facultas Scientiae."[22] Or, in English, "Science enables Exploration, Exploration enables Science." This is the kind of positive feedback that gets things done. Now, thanks to the rapid changes happening in space technology, we can extend his idea: "Scientia facultas Cupiditatis, Cupiditas facultas Scientiae"—"Science enables Greed, Greed enables Science." This virtuous cycle can solve the simultaneous crises that astronomy and planetary science face and will let us remove the threat of being wiped out by an asteroid.

Greed will lead us to work in space on a whole new scale. Profits from space will lower the costs of doing anything in space, including science. If space mining reached the scale of Walmart, then just 1 percent of its revenue applied to prospecting would be nearly double NASA's total current budget. Also, the pressure to make a profit will push down the prices for spacecraft and launches so we can get more bang for our research buck. The ability to run complex mining operations in space will translate into being able not just to deflect asteroids bound on a collision course with Earth, but also to turn them into habitats. Our greater powers to build in space will let us create telescopes and planetary probes so big that they are now just a fantasy. That will bring exploration of the Solar System to scale, including human exploration, and give us a new golden—or perhaps it should be platinum—age of exploration of our Solar System and the universe.

Fear will be allayed by greed. Making space part of the economy by leveraging greed, we can deal with our fear. We can completely retire the risk of a Tunguska-like impact wiping out Washington, DC, or of the Big One destroying our species. To be able to deflect an asteroid heading for Earth today means we need to be able to find it while it is still far away, so we have enough time to prepare a whole new deflection mission from scratch. Instead, when we have a thriving in-space business of mining the asteroids, we can rapidly respond to a newly found threatening asteroid, so having a long warning time is not so crucial. We will have powerful rockets continually going back and forth to asteroids, so all we will need to do is change the payload of one or two of them from mining equipment to a nuclear device. If the rockets are powerful enough, we may not even need that; a heavy chunk of metal will be enough to deflect the killer rock. We will get longer warning times too. There will be many companies scouring the skies for all the asteroids that are out there to find the best ones to mine. We will have to make it in their interests to reveal any hazardous asteroids they find. For instance, they could be made liable for damages if they don't let us know in good time. That should be incentive enough; cities are quite expensive.

And what of love? How will love's frustration be requited? Will our curiosity, our love of finding things out, be helped by greed? Yes. Our curiosity is constrained by cost. What might we do if greed makes a vastly expanded exploration of the universe affordable? Having infrastructure established in deep space will yield great benefits for scientific missions too. First let's look at exploring the Solar System.

Famously, to business Time is Money, so advanced rockets to bring the valuable ores back quickly will be built to maximize profit. These could be solar powered in the inner Solar System, maybe as far as the Main Belt. But further out, in the realm of the gas giant planets and their many moons, sunlight is reduced enough that nuclear-powered rockets will become the clear favorites. If we get fusion-powered rockets, then we can have direct trips to Pluto that take 4 years instead of the almost 10 years it took *New Horizons* in its looping path around the planets to gain speed. That speed turns a mission to the outer planets into one episode in a career, instead of a whole career. Psychologically, more people will surely be motivated to join efforts if the rewards come sooner. There would be enough power from a fusion rocket, if we can harness it, to slow down and go into orbit around Pluto or other distant worlds, including the putative Planet 9, if it exists. Orbiting any planet pours back far more information than a simple fly-by. And because of the commercial imperative to cut cost, the missions will be so much cheaper that we can have a bunch on the go at once.

Faster trips mean less radiation damage to astronauts too. Humans are much more efficient explorers than robotic craft. Steve Squyres, the principal investigator of the *Spirit* and *Opportunity* Mars rovers, wrote in his 2005 book *Roving Mars,* "The unfortunate truth is that most things our rovers can do in a perfect sol [a Martian day, which is 24 hours, 37 minutes long] a human explorer could do in less than a minute."[23] Robotics is improving fast, and space mining will provide a good training ground for AI algorithms to learn about exploring space. Then we will get better value out of our future robotic explorers. At the same time space mining will make human space travel easier, safer, and quicker. We will then have more chances to deploy our own talents for exploration in the field, as geologists do today.

Moving further out there are other solar systems, "exoplanets." The big question is whether they teem with life, like a second Earth, or are barren. If we could show that 99 percent of them are dead worlds, that would make our home world extraordinarily precious. Knowing that would give our small selves a truly cosmic significance, as Martin Rees, the UK astronomer royal, puts it.[24] For a while NASA was studying in earnest a "Terrestrial Planet Finder," a mission big and powerful enough to see Earth-like planets in the "habitable zone" around hundreds of nearby stars, if such planets exist.[25] Dreams of a large telescope in space were abandoned. Work has restarted on the idea, but it looks as though this one telescope would cost twice as much as NASA has to spend in 10 years, leaving nothing left over for anything else. Supporters argue that this is worth it, and that answering the Big Question "Are we alone?" would justify a big boost to NASA's astrophysics budget. That would be great. If it doesn't happen, though, this is a dream that space mining could make real. Space mining will grow our abilities to build things in space while simultaneously slicing the cost to a fraction of what it would currently take. A truly large telescope equipped with a "starshade" to cut out the glare from the exo-Earth's Sun that could survey hundreds of candidates for "Earth 2.0" would then become affordable.

If we find another Earth with signs of life, could we actually take a picture of it that was detailed enough to see clouds, continents, oceans, and maybe more? There is a location 550 times further from the Sun than the Earth is, where the Sun focuses light from behind it just by its warping of space. It is called the solar gravitational lens focus.[26] If we put a telescope out there, directly opposite the Sun to where another Earth is, then we'll see far more detail than just continents. Getting out there is the big problem. It is 10 times more distant even than Pluto. So fast rockets will be essential. That's not the only challenge. The image it forms is huge, so we'll need some clever way of capturing that. And the image moves as the Other Earth orbits its own Sun, so our giant telescope will have to keep moving too. With today's technology these are nigh on impossible challenges. But space mining could render them tractable.

The most direct method of taking images of exo-Earths is to send spacecraft with cameras to the stars. Breakthrough Starshot is a dream of billionaire Yuri Milner, Harvard astronomer Avi Loeb, and others to send fleets of tiny spaceships, each no bigger than a credit card, to Proxima

Centauri, our nearest stellar neighbor, four light-years away.[27] Those nano-satellites that survive the trip will fly through the Proxima Centauri system and send back a few snapshots each. They should capture enough detail to see oceans and continents on planets in the system, if they exist. They would be propelled by a giant gigawatt laser using as much power as an Earth city of a quarter-million people. As a powerful laser firing from the ground could also be used to destroy satellites in orbit, some may get nervous at that prospect. There is only one place in the Solar System that can never see those satellites—the far side of the Moon. A giant installation there is unlikely to happen without a well-developed space economy.

I can't help wondering what we would think if the Proxima Centaurians had their own Breakthrough Starshot that sent hundreds of little spy craft through our Solar System with a powerful laser. Might we get nervous about their intent? Perhaps (spoiler alert) I have just taken too much to heart science fiction writer Cixin Liu's *Dark Forest* vision of a universe full of hostile life?

Going lastly to the scale of the universe, what might we accomplish? The far side of the Moon not only blocks a giant laser there from doing damage to Earth-orbiting satellites, it is also shielded from all the radio noise generated on Earth. It is a cosmic "radio quiet zone." As such it is the perfect place to install a radio telescope to listen for the faint signals from hydrogen atoms before the epoch of "Cosmic Dawn," when the first stars formed. We can make a start on this now by checking out just how quiet the far side really is and by looking for a good site. But to build the full-up telescope some 200 kilometers across will need the abilities and cost savings that space mining, in this case lunar mining, will provide.

In the 400 years since Galileo first looked at the sky with a telescope, our telescopes have advanced so much that we can see stars billions of times fainter than he could discern. Yet those mighty telescopes make images that are only about 1,000 times sharper than Galileo could see. That's a lot, but pales before our far greater advance in how deep we can see into the universe compared with him. Yet every time we do gain a little in image sharpness, we learn an enormous amount. It's like looking the answer up in the back of the book. But the sizes of our telescopes are now limited, and that sets a fundamental limit to how sharp their images can be.

The key to making better images is a technique called interferometry. We don't have to have a single giant mirror but can use a number of smaller telescopes spread over large distances to make sharper images. NASA recognizes this. A 2013 report led by Chryssa Kouveliotou of George Washington University set out its vision for the 30-year future of astronomy.[28] This vision was built around using interferometers to get radically better angular resolution. There are a few hints at what that could give us. CHARA, a relatively modest experimental interferometer run by Georgia State University and located on Mount Wilson in California, has made pictures of stars that show some to be flattened into oblongs due to their fast rotation, and others that have giant starspots.[29] A more ambitious new instrument on the European Very Large Telescope Interferometer has just started to make out the structures around giant black holes.[30]

The sharpest images we have today come from radio interferometry. With this technique astronomers have imaged the shadow of a black hole using the Event Horizon Telescope.[31] But only just. There's a long way to go with improving those images. From the ground we cannot get to see sharper detail. The Earth is just not big enough. To get a sharper image we must launch a suite of radio telescopes into space and spread them out over many times the size of the Earth yet link them up to work together. Then we could see a huge amount of detail from the hot plasma about to disappear forever inside the black hole. We also know of ways to make ultra-high-resolution images of black holes in gamma rays and X-rays too. It will take a lot of effort to make these telescopes a reality, placing clever lenses thousands of kilometers from the detectors. We can't afford to build them right now, but as space mining will make them cheaper and easier to build, the cost-benefit will get vastly better. They are in our future.

I hope that you now agree that by using greed to harness the vast resources of space we can assuage our fear and requite our love of knowledge. What will be the result? To paraphrase Shakespeare's King Lear:

> We shall do such things—
> What they are, yet we know not: but they shall be
> The wonders of the Earth.[32]

And not just the Earth.

Notes

Introduction

1. As space entrepreneur Peter Diamandis asked in 2008: "Why do we explore space? . . . There are three drivers, Fear, Curiosity, and Wealth." His next statement was met with laughter: "You can easily measure the ratio of fear to curiosity as the ratio of the Defense budget to the science budget." "Singularity Summit—Enlightened Machines, Plus an Exclusive Interview with Peter Diamandis," *Valley Zen* (blog), 12 November 2008, http://www.valleyzen.com/2008/11/12/singularity-summit-enlightened-machines-interview-peter-diamandis/.

2. *The Stars Are Indifferent to Astronomy* is the title of Nada Surf's 2011 album. They radiate where they will, heedless of our technological limitations. http://www.nadasurf.com/albums/the-stars-are-indifferent-to-astronomy/ (accessed 10 July 2020).

3. National Research Council of the National Academies, *Vision and Voyages for Planetary Science in the Decade 2013–2022* (Washington, DC: National Academies Press, 2011), https://doi.org/10.17226/13117.

1. Asteroids

1. Iron: density = 7.87 ("Iron," Wikipedia, https://en.wikipedia.org/wiki/Iron [accessed 10 July 2020]); Silicon: density = 2.33 ("Silicon," Wikipedia, https://en.wikipedia.org/wiki/Silicon [accessed 10 July 2020]).

2. W. Herschel, "Observations on the Two Lately Discovered Celestial Bodies," *Philosophical Transactions of the Royal Society of London* 92 (1802): 213–232. There has been some dispute about who came up with the name, but it appears settled in Herschel's favor: C. Cunningham and W. Orchinson, "Who Invented the Word Asteroid: William Herschel or Stephen Weston?" *Journal of Astronomical History and Heritage* 14, no. 3 (2011): 230–34.

3. Herschel, "Observations on the Two Lately Discovered Celestial Bodies."

4. They include quasar, active galactic nucleus (AGN), Seyfert galaxy, X-ray bright/optically normal galaxy (XBONG), optically dull AGN, low ionization nuclear emission line region (LINER), type 1 AGN, type 2 AGN, types 1.5, 1.8, and 1.9 AGN, narrow line Seyfert 1, narrow emission line galaxy, dust obscured galaxies (DOGs), Compton thin AGN, Compton thick AGN, true/naked type 2 AGN, changing look AGN, quasi-stellar object, radio loud quasar, radio galaxy, Fanaroff-Riley type I radio galaxy, Fanaroff-Riley type II radio galaxy, blazar, BL lac object, high frequency peaked blazar, low frequency peaked blazar, intermediate frequency peaked blazar, flat spectrum radio galaxy, compact symmetric source, and gigahertz peaked source.

5. "Some Scientific Centres. IV—The Heidelberg Physical Laboratory," *Nature* 65 (1902): 587.

6. The very same year that biology was transformed too, when Charles Darwin published *On the Origin of Species,* introducing the idea of evolution by natural selection.

7. G. Kirchhoff, "On the Relation between the Radiating and Absorbing Powers of Different Bodies for Light and Heat," *Philosophical Magazine,* ser. 4, 20 (1860): 1–21.

8. G. Beekman, "The Nearly Forgotten Scientist Ivan Osipovich Yarkovsky," *Journal of the British Astronomical Association* 115, no. 4 (2005): 207.

9. See figure 7 in A. Taylor, J. C. McDowell, and M. Elvis, "Phobos and Mars Orbit as a Base for Asteroid Mining," *Acta Astronautica* (submitted).

10. As recounted by the collaboration of a meteoriticist, Diane Johnson, and an Egyptologist, Joyce Tyldesley. D. Johnson and J. Tyldesley, "Iron from the Sky," *Geoscientist Online,* 2014, https://www.geolsoc.org.uk/Geoscientist/Archive/April-2014/Iron-from-the-sky.

11. Cathryn J. Prince, *A Professor, a President, and a Meteor: The Birth of American Science* (New York: Prometheus, 2011).

12. Harry Y. McSween, "Ensisheim Meteorite," Brittanica.com, https://www.britannica.com/topic/Ensisheim-meteorite.

13. L. R. Nittler and N. Dauphas, "Meteorites and the Chemical Evolution of the Milky Way," in *Meteorites and the Early Solar System II*, ed. D. S. Lauretta and H. Y. McSween Jr. (Tucson: University of Arizona, 2006), 127–46.

2. Love

1. Richard B. Setlow, "The Hazards of Space Travel," *EMBO Reports* 4, no. 11 (2003): 1013, doi:10.1038/sj.embor.7400016.

2. Stewart Weaver, *Exploration: A Very Short Introduction* (Oxford: Oxford University Press, 2015).

3. Ben R. Finney, "Exploring and Settling Pacific Ocean Space—Past Analogues for Future Events?" in *Space Manufacturing 4: Proceedings of the Fifth Princeton/AIAA Conference, May 18–21, 1981,* ed. Jerry Grey and Lawrence A. Hamdan (New York: American Institute of Aeronautics and Astronautics, 1981), 261; Ben R. Finney, *From Sea to Space* (Palmerston North, NZ: Massey University, 1992).

4. M. Lecar, M. Podolak, D. Sasselov, and E. Chiang, "On the Location of the Snow Line in a Protoplanetary Disk," *Astrophysical Journal* 640 (2006): 1115.

5. R. Gomes, H. F. Levison, K. Tsiganis, and A. Morbidelli, "Origin of the Cataclysmic Late Heavy Bombardment Period of the Terrestrial Planets," *Nature* 435, no. 7041 (2005): 466–69, doi:10.1038/nature03676; K. Tsiganis, R. Gomes, A. Morbidelli, and H. F. Levison, "Origin of the Orbital Architecture of the Giant Planets of the Solar System," *Nature* 435, no. 7041 (2005): 459–61, doi:10.1038/nature03539; A. Morbidelli, H. F. Levison, K. Tsiganis, and R. Gomes, "Chaotic Capture of Jupiter's Trojan Asteroids in the Early Solar System," *Nature* 435, no. 7041 (2005): 462–65, doi:10.1038/nature03540.

6. Most probably not from the *Analects* of Confucius, but from a Chinese fable: Annie Feng, "What is the meaning of 'the man who moves a mountain begins by carrying stones'?" Goodreads, https://www.goodreads.com/questions/1156657-what-is-the-meaning-of-the-man-who-moves/answers/625580-i-m-working-through (accessed 3 September 2020).

7. T. Encrenaz, "Water in the Solar System," *Annual Reviews of Astronomy & Astrophysics* 46 (2008): 57, available at https://www-annualreviews-org.ezpprod1.hul.harvard.edu/doi/pdf/10.1146/annurev.astro.46.060407.145229; L. M. Prockter, "Ice in the Solar System," *Johns Hopkins APL Technical Digest* 26, no. 2 (2005), https://www.jhuapl.edu/techdigest/TD/td2602/Prockter.pdf.

8. Sara Schechner, *Comets, Popular Culture and the Birth of Modern Cosmology* (Princeton, NJ: Princeton University Press, 1997).

9. B. W. Eakins and G. F. Sherman, "Volumes of the World's Oceans from ETOPO1," NOAA National Geophysical Data Center, 2010, https://ngdc.noaa.gov/mgg/global/etopo1_ocean_volumes.html.

10. T. H. Prettyman et al., "Extensive Water Ice within Ceres' Aqueously Altered Regolith: Evidence from Nuclear Spectroscopy," *Science* 355 (2017): 55.

11. W. F. Bottke, R. J. Walker, J. M. D. Day, D. Nesvorny, and L. Elkins-Tanton, "Stochastic Late Accretion to Earth, the Moon, and Mars," *Science* 330 (2010): 1527.

12. Title of a 1990 NASA *Voyager 1* photo of the Earth from Saturn, taken at the request of Carl Sagan, and the title of one of his books. " 'Pale Blue Dot' Images Turn 25," NASA, 11 February 2015, https://www.nasa.gov/jpl/voyager/pale-blue-dot-images-turn-25; "Voyager 1's Pale Blue Dot," NASA, https://solarsystem.nasa.gov/resources/536/voyager-1s-pale-blue-dot (accessed 3 September 2020); Carl Sagan, *Pale Blue Dot: A Vision of the Human Future in Space* (New York: Random House, 1994).

13. Compared with other planets, the Earth's atmosphere is blue because oxygen destroyed the brown organic compounds. A new result says that the ocean is blue because of its own color, especially when seen from overhead. NASA Share the Science, "Ocean Color," https://science.nasa.gov/earth-science/oceanography/living-ocean/ocean-color (accessed 16 January 2021).

14. See, for example, the MIT journal *Artificial Life* at https://www.mitpressjournals.org/loi/artl.

15. C. Chyba and C. Sagan, "Endogenous Production, Exogenous Delivery and Impact-Shock Synthesis of Organic Molecules: An Inventory for the Origins of Life," *Nature* (1992): 355, 125.

16. E. Herbst and E. F. van Dishoeck, "Complex Organic Interstellar Molecules," *Annual Reviews of Astronomy and Astrophysics* 47 (2009): 427–80; J. S. Carr and J. R. Najita, "Organic Molecules and Water in the Inner Disks of T Tauri Stars," *Astrophysical Journal* 733, no. 2 (2011): 102.

17. Julie E. M. McGeoch and Malcolm W. McGeoch, "Polymer Amide in the Allende and Murchison Meteorites," *Meteoritics & Planetary Science* 50, no. 12 (2015): 1971–83.

18. Jane Gregory, *Fred Hoyle's Universe* (Oxford: Oxford University Press, 2005), 324.

19. Gregory, *Fred Hoyle's Universe*, 283; Sir Fred Hoyle, "Comets—A Matter of Life and Death," *Vistas in Astronomy* 24 (1980): 123, https://www.sciencedirect.com/science/article/abs/pii/0083665680900276; F. Hoyle, *The Relation of Astronomy to Biology* (Cardiff: University College Cardiff Press, 1980).

20. Keith Kvenvolden, James Lawless, Katherine Pering, Etta Peterson, Jose Flores, Cyril Ponnamperuma, I. R. Kaplan, and Carleton Moore, "Evidence for Extraterrestrial Amino-Acids and Hydrocarbons in the Murchison Meteorite," *Nature* 228 (1970): 923–26, https://www.nature.com/articles/228923a0.

20. D. S. McKay et al., "Search for Past Life on Mars: Possible Relic Biogenic Activity in Martian Meteorite ALH84001," *Science* 273 (1996): 924–30.

22. K. Meach et al., "A Brief Visit from a Red and Extremely Elongated Interstellar Asteroid," *Nature* 552 (2017): 378; Amir Siraj and Avi Loeb, "Discovery of a Meteor of Interstellar Origin," 2019, arXiv:1904.07224, https://arxiv.org/pdf/1904.07224.pdf.

23. Matthias Willbold, Tim Elliott, and Stephen Moorbath, "The Tungsten Isotopic Composition of the Earth's Mantle before the Terminal Bombardment," *Nature* 477 (2011): 195.

24. The Isua supracrustal belt is a small patch of exposed rock 30 meters by 70 meters that is dated to be 3.7–3.8 billion years old. This is the same place that the earliest fossils may have been found. Allen P. Nutman, Vickie C. Bennett, Clark R. L. Friend, Martin J. Van Kranendonk, and Allan R. Chivas, "Rapid Emergence of Life Shown by Discovery of 3,700-Million-Year-Old Microbial Structures," *Nature* 537 (2016): 535–38; Carolyn Gramling, "Hints of Oldest Fossil Life Found in Greenland Rocks," *Science*, 31 August 2016, http://www.sciencemag.org/news/2016/08/hints-oldest-fossil-life-found-greenland-rocks; Nicholas Wade, "World's Oldest Fossils Found in Greenland," *New York Times*, 31 August 2016, http://www.nytimes.com/2016/09/01/science/oldest-fossils-on-earth.html?_r=0.

25. Mario Fischer-Gödde, Bo-Magnus Elfers, Carsten Münker, Kristoffer Szilas, Wolfgang D. Maier, Nils Messling, Tomoaki Morishita, Martin Van Kranendonk, and Hugh Smithies, "Ruthenium Isotope Vestige of Earth's Pre-Late-Veneer Mantle Preserved in Archaean Rocks," *Nature* 579 (2020): 240, doi.org/10.1038/s41586-020-2069-3.

3. Fear

1. Terry Bisson, "Meat," Terry Bisson of the Universe, http://www.terrybisson.com/theyre-made-out-of-meat-2/ (accessed 4 September 2020).

2. Donald E. Osterbrock, *Walter Baade: A Life in Astrophysics* (Princeton, NJ: Princeton University Press, 2002).

3. Philip Plait, *Death from the Skies: "These are the ways the world will end . . . ,"* (New York: Viking Penguin, 2008).

4. D. Kring and M. Boslough, "Chelyabinsk: Portrait of an Asteroid Airburst," *Physics Today* 67, no. 9 (2014): 32, https://doi.org/10.1063/PT.3.2515.

5. This is not the only case of people's curiosity beating their fear. The app Citizen (https://citizen.com) was designed to alert folks to nearby dangers so that they could avoid them; instead, people went over to look! *Wait Wait, Don't Tell Me*, NPR, 16 March 2019, https://www.npr.org/2019/03/16/704081183/panel-questions.

6. A. P. Kartashova, O. P. Popova, D. O. Glazachev, P. Jenniskens, V. V. Emel'yanenko, E. D. Podobnaya, and A. Ya Skripnik, "Study of Injuries from the Chelyabinsk Airburst Event," *Planetary & Space Science* 160 (2018): 107–14.

7. Stefan Geens, "Reconstructing the Chelyabinsk Meteor's Path, with Google Earth, YouTube and High-School Math," *Ogle Earth* (blog), 16 February 2013, http://goo.gl/vcG3Y.

8. Jorge I. Zuluaga and Ignacio Ferrin, "A Preliminary Reconstruction of the Orbit of the Chelyabinsk Meteoroid," 2013, arXiv:1302.5377. For another confirmation, see S. R. Proud, "Reconstructing the Orbit of the Chelyabinsk Meteor Using Satellite Observations," *Geophysical Research Letters* 40, no. 13 (2013): 3351–55. https://doi.org/10.1002/grl.50660.

9. A. W. Harris, "What Spaceguard Did," *Nature* 453 (2008): 1178–79; P. G. Brown et al., "A 500-Kiloton Airburst over Chelyabinsk and an Enhanced Hazard from Small Impactors," *Nature* 503 (2013): 238–41.

10. Leonard David, "Huge Meteor Explosion a Wake-Up Call for Planetary Defense," *Scientific American,* 21 March 2019, https://www.scientificamerican.com/article/huge-meteor-explosion-a-wake-up-call-for-planetary-defense/.

11. Lindley Johnson, presentation to the 17th Meeting of the NASA Small Bodies Advisory Group, June 2017, https://www.lpi.usra.edu/sbag/meetings/jun2017/.

12. Alan W. Harris and Germano D'Abramo, "The Population of Near-Earth Asteroids," *Icarus* 257 (2015): 302–12.

13. David Morrison, "Tunguska Workshop: Applying Modern Tools to Understand the 1908 Tunguska Impact" (NASA Technical Memorandum 220174, 2018), NASA Scientific and Technical Information (STI) Program; Paolo Farinella, L. Foschini, Christiane Froeschlé, R. Gonczi, T. J. Jopek, G. Longo, and Patrick Michel, "Probable Asteroidal Origin of the Tunguska Cosmic Body," *Astronomy & Astrophysics* 377, no. 3 (2001): 1081–97, doi:10.1051/0004-6361:20011054.

14. M. Boslough, "Computational Modeling of Low-Altitude Airbursts," American Geophysical Union, Fall Meeting 2007, abstract id. U21E-03.

15. Planetary and Space Science Centre (PASSC), Earth Impact Database, http://www.passc.net/EarthImpactDatabase/index.html.

16. H. St Jean Philby, *The Empty Quarter* (New York: Henry Holt, 1933); J. C. Wynn and E. M. Shoemaker, "The Day the Sands Caught Fire," *Scientific American* 279, no. 5 (1998): 36–45; E. M. Shoemaker and J. C. Wynn, "Geology of the Wabar Meteorite Craters, Saudi Arabia" (abstract), *Lunar and Planetary Science* 28 (1997): 1313–14; Jeff Wynn and Gene Shoemaker, "The Wabar Meteorite Impact Site, Ar-Rub'

Al-Khali Desert, Saudi Arabia," U.S. Geological Survey, https://volcanoes.usgs.gov/jwynn/3wabar.html (accessed 4 September 2020).

17. H. M. Basurah, "Estimating a New Date for the Wabar Meteorite Impact," *Meteoritics & Planetary Science* 38, no. 7 (2003), supplement, A155–56.

18. The most recent known craters on Earth are:

Name	Age (years)	Diameter (meters)
Carancas	12*	13.5
Sikhote Alin	73*	20
Wabar	140**	110
Haviland	<1,000	10
Sobolev	<1,000	50

* Dating from 2020.
** 290±38 years of age according to J. Prescott, G. Robertson, C. Shoemaker, E. M. Shoemaker, and J. C. Wynn, "Luminescence Dating of the Wabar Meteorite Craters, Saudi Arabia," *Journal of Geophysical Research* 109 (2004): E01008, doi:10.1029/2003JE002136.

Source: PASSC, Earth Impact Database, http://www.passc.net/EarthImpactDatabase/New%20website_05-2018/Agesort.html.

19. David A. Kring, *Guidebook to the Geology of Barringer Meteorite Crater, Arizona (a.k.a. Meteor Crater)*, 2nd ed., LPI Contribution No. 2040, prepared for the 80th Annual Meeting of the Meteoritical Society, July 2017, Lunar and Planetary Institute, https://www.lpi.usra.edu/publications/books/barringer_crater_guidebook/.

20. Yasunobu Uchiyama, Felix A. Aharonian, Takaaki Tanaka, Tadayuki Takahashi, and Yoshitomo Maeda, "Extremely Fast Acceleration of Cosmic Rays in a Supernova Remnant," *Nature* 449 (2007): 576; "NASA: Major Step toward Knowing Origin of Cosmic Rays," NASA press release 07-63, 9 October 2007, https://www.nasa.gov/centers/goddard/news/topstory/2007/accelerated_rays.html.

21. Prescott et al., "Luminescence Dating of the Wabar Meteorite Craters, Saudi Arabia."

22. Eugene M. Shoemaker, "Impact Mechanics at Meteor Crater, Arizona," in *The Moon Meteorites and Comets*, ed. Barbara Middlehurst and Gerard P. Kuiper (Chicago: University of Chicago Press, 1963), 301, http://articles.adsabs.harvard.edu/pdf/1963mmc..book..301S.

23. D. A. Kring, "Air Blast Produced by the Meteor Crater Impact Event and a Reconstruction of the Affected Environment," *Meteoritics & Planetary Science* 32 (1997): 517–30.

24. Daniel J. Field et al., "Early Evolution of Modern Birds Structured by Global Forest Collapse at the End-Cretaceous Mass Extinction," *Current Biology* 28 (2018): 1825–31, https://doi.org/10.1016/j.cub.2018.04.062.

25. Derek W. Larson, Caleb M. Brown, and David C. Evans, "Dental Disparity and Ecological Stability in Bird-like Dinosaurs prior to the End-Cretaceous Mass Extinction," *Current Biology* 26 (2016): 1325–33, https://doi.org/10.1016/j.cub.2016.03.039; "Fossil Teeth Suggest That Seeds Saved Bird Ancestors from Extinc-

tion," Phys.org, 16 April 2016, https://phys.org/news/2016-04-fossil-teeth-seeds-bird-ancestors.html.

26. So how did mammals survive? We all have teeth, not beaks. Perhaps there were some great hoarders among our ancestors, an extreme version of squirrels, who had saved up large underground caches of seeds? No doubt someone will find a way to determine what happened.

27. See the list of dinosaur death theories at the University of California Museum of Paleontology: "What Killed the Dinosaurs?" DinoBuzz, http://www.ucmp.berkeley.edu/diapsids/extinction.html (accessed 5 September 2020).

28. Bianca Bosker, "The Nastiest Feud in Science," *Atlantic*, September 2018, https://www.theatlantic.com/magazine/archive/2018/09/dinosaur-extinction-debate/565769/.

29. For example, Gerta Keller, "Cretaceous Climate, Volcanism, Impacts, and Biotic Effects," *Cretaceous Research* 29 (2008): 754–71, doi.org/10.1016/j.cretres.2008.05.030. See also Professor Keller's website on Deccan Volcanism: doi.org/10.1016/j.cretres.2008.05.030.

30. Paul Voosen, "Did Volcanic Eruptions Help Kill off the Dinosaurs?" Sciencemag.org, 21 February 2019, doi:10.1126/science.aax1020.

31. H.R. Report 1022, George E. Brown, Jr., Near-Earth Object Survey Act, 109th Cong. (2005–6), www.GovTrack.us, https://www.govtrack.us/congress/bills/109/hr1022; NASA Near-Earth Object Observations Program: https://www.nasa.gov/planetarydefense/neoo.

32. Harris and D'Abramo, "The Population of Near-Earth Asteroids."

33. Brown et al., "A 500-Kiloton Airburst over Chelyabinsk," 238. Peter Brown's website: http://www.physics.uwo.ca/people/faculty_web_pages/brown.html.

34. Harris, "What Spaceguard Did."

35. See chapter 7, "A Low Risk, but Not Negligible," in Martin Rees, *Our Final Hour* (New York: Basic Books, 2003).

36. *Make a Dent in the Universe: Erika Ilves at TEDxStavanger* (video), YouTube, 25 September 2013, https://www.youtube.com/watch?v=K89nP7GhWgU.

37. *Planetary Defense Conference Exercise—2015*, NASA Center for Near-Earth Object Studies (CNEOS), https://cneos.jpl.nasa.gov/pd/cs/pdc15/. Each day has its own web page.

38. *Planetary Defense Conference Exercise—2019*, NASA Center for Near-Earth Object Studies (CNEOS), https://cneos.jpl.nasa.gov/pd/cs/pdc19/. Each day has its own web page.

39. International Asteroid Warning Network (IAWN): http://iawn.net/about.shtml.

40. Space Mission Planning Advisory Group: https://www.cosmos.esa.int/web/smpag/home.

41. "Astronomers Complete First International Asteroid Tracking Exercise," Jet Propulsion Laboratory, 3 November 2017, https://www.jpl.nasa.gov/news/news.php?feature=6994.

42. Vishnu Reddy et al., "Near-Earth Asteroid 2012 TC4 Observing Campaign: Results from a Global Planetary Defense Exercise," *Icarus* (2019): 133–50.

4. Greed

1. "Peter Diamandis: The First Trillionaire Is Going to Be Made in Space," *Business Insider,* 2 March 2015, https://www.businessinsider.com/peter-diamandis-space-trillionaire-entrepreneur-2015-2?IR=T&r=SG.

2. Bryan Bender, "Ted Cruz: 'The First Trillionaire Will Be Made in Space,'" *Politico,* 1 June 2018, https://www.politico.com/story/2018/06/01/ted-cruz-space-first-trillionaire-616314; Katie Kramer, "Neil deGrasse Tyson Says Space Ventures Will Spawn First Trillionaire," *NBC News,* 3 May 2015, https://www.nbcnews.com/sci ence/space/neil-degrasse-tyson-says-space-ventures-will-spawn-first-trillion-aire-n352271.

3. Government Accounting Office, *NASA: Constellation Program Cost and Schedule Will Remain Uncertain Until a Sound Business Case Is Established,* GAO-09-844, 26 August, 2009, publicly released 25 September 2009, https://www.gao.gov/prod ucts/GAO-09-844.

4. John S. Lewis, *Mining the Sky: Untold Riches from the Asteroids, Comets, and Planets* (Reading, MA: Addison-Wesley, 1996).

5. "In space, no one can hear you scream." From the 1979 movie *Alien,* directed by Ridley Scott and starring Sigourney Weaver.

6. Sixty trillion UK pounds. Rob Waugh, "Single Asteroid Worth £60 Trillion if It Was Mined—as Much as World Earns in a Year," *Daily Mail,* 21 May 2012. (The tabloid the *Daily Mail* has good reporting on space.) The GDP of the entire world for 2012 was only about U.S.$75 trillion; see World Bank: https://data.worldbank.org/ indicator/NY.GDP.MKTP.CD.

7. Asterank.com (accessed 14 July 2020). Unfortunately, 9283 Martinelvis has an assigned value of $0.

8. Jonas Peter Akins, "Short Lays on Greasy Voyages: Whaling and Venture Capital," Nantucket Historical Association, July 2018, https://nha.org/research/ nantucket-history/history-topics/short-lays-on-greasy-voyages-whaling-and-venture-capital/.

9. "Platinum Fact Sheet," AZO Materials, *AzoM,* 11 February 2002, https://www. azom.com/article.aspx?ArticleID=1238.

10. Abundances in the Earth's crust: Pt: 0.005 mg/kg (=ppm); Au: 0.004; Pd: 0.015. "Abundance of Elements in the Earth's Crust and in the Sea," *CRC Handbook of Chemistry and Physics,* ed. William M. Haynes, 97th ed. (Boca Raton, FL: CRC, 2016–17), 14–17. (Note for astronomers: these are not solar abundances.)

11. R. Grant Cawthorn, "Seventy-Fifth Anniversary of the Discovery of the Platiniferous Merensky Reef: The Largest Platinum Deposits in the World," *Platinum Metals Review,* 43, no. 4 (1999): 146.

12. Kelly Weinersmith and Zach Weinersmith, *Soonish: Ten Emerging Technologies That'll Improve and/or Ruin Everything* (New York: Penguin, 2017).

13. David K. Israel, "5 of the Most Expensive Bottles of Wine Ever Sold," *Week,* 7 November 2013, http://theweek.com/articles/457193/5-most-expensive-bottles-wine-ever-sold. Caviar is not in the competition, by the way. Ossetra caviar retails at only about $3 per gram.

14. Ed Caesar, "The Woman Shaking Up the Diamond Industry," *New Yorker,* 27 January 2020, https://www.newyorker.com/magazine/2020/02/03/the-woman-shaking-up-the-diamond-industry.

15. Caleb Henry, "Northrop Grumman's *MEV-1* Servicer Docks with Intelsat Satellite," *Space News,* 26 February 2020.

16. Donald J. Kessler and Burton G. Cour-Palais, "Collision Frequency of Artificial Satellites: The Creation of a Debris Belt," *Journal of Geophysical Research* 83, no. A6 (1978): 2637–46.

17. Jonathan's Space Report: planet4589.org.

18. *RemoveDEBRIS,* Surrey Space Centre, University of Surrey, https://www.surrey.ac.uk/surrey-space-centre/missions/removedebris (accessed 4 September 2020).

19. Emre Kelly, "NASA Shows Interest in SpaceX's Starship Orbital Refueling Ambition," *Florida Today,* 12 October 2019, https://www.floridatoday.com/story/tech/science/space/2019/10/12/nasa-shows-interest-spacexs-starship-orbital-refueling-ambitions/3957775002/.

20. Alessondra Springmann et al., "Thermal Alteration of Labile Elements in Carbonaceous Chondrites," *Icarus* 324 (2019): 104.

21. Evan M. Melhado, "Jöns Jacob Berzelius, Swedish Chemist," last updated 16 August 2020, *Encyclopedia Brittanica,* https://www.britannica.com/biography/Jons-Jacob-Berzelius.

22. J. O'Neil et al., "The Nuvvagittuq Greenstone Belt," in M. Van Kranendonk, V. Bennett, and E. Hoffmann, eds., *Earth's Oldest Rocks* (Amsterdam: Elsevier, 2007), 349–374. doi: 10.1016/B978-0-444-63901-1.00016-2.

23. "Widmanstätten Pattern, Astronomy," last updated 9 April 2012, *Encyclopaedia Britannica,* https://www.britannica.com/science/Widmanstatten-pattern.

24. N. L. Hooper, "Space Rocks: A Series of Papers on Meteorites and Asteroids" (senior thesis, Harvard University, 2016).

25. Kevin Cannon, private communication using data from the Meteoritical Bulletin Database, https://www.lpi.usra.edu/meteor/, 2018. My thanks to him for compiling these numbers.

26. James Scott Berdahl, "Morning Light: The Secret History of the Tagish Lake Fireball" (MS thesis, Massachusetts Institute of Technology, 2010), https://cmsw.mit.edu/wp/wp-content/uploads/2016/06/227233756-James-Berdahl-Morning-Light-The-Secret-History-of-the-Tagish-Lake-Fireball.pdf.

27. "First Detection of Sugars in Meteorites Gives Clues to Origin of Life," NASA press release 19-23, 18 November 2019, https://www.nasa.gov/press-release/goddard/2019/sugars-in-meteorites; Yoshihiro Furukawa et al., "Extraterrestrial Ribose and Other Sugars in Primitive Meteorites," *Proceedings of the National Academy of Sciences* 116, no. 49 (2019): 24440–45, https://doi.org/10.1073/pnas.1907169116.

28. B. A. Macleod, J. T. Wang, C. C. Chung, C. R. Ries, S. K. Schwarz, and E. Puil, "Analgesic Properties of the Novel Amino Acid, Isovaline," *Anesthesia & Analgesia* 110, no. 4 (2010): 1206–14, doi:10.1213/ane.0b013e3181d27da2, PMID 20357156.

29. Paul Steinhardt has recounted the story of his discovery of quasicrystals in meteorites in *The Second Kind of Impossible: The Extraordinary Quest for a New Form of Matter* (New York: Simon & Schuster, 2019). A shorter account is given in L. Bindi and

P. J. Steinhardt, "The Quest for Forbidden Crystals," *Mineralogical Magazine* 78, no. 2 (2014): 467–82, http://www.physics.princeton.edu/~steinh/lucaMinMag.pdf, and in Paul J. Steinhardt and Luca Bindi, "Once upon a Time in Kamchatka: The Search for Natural Quasicrystals," http://physics.princeton.edu/~steinh/SteinhardtICQ11r.pdf. There are many links to the original academic papers at Paul Steinhardt's website: http://www.physics.princeton.edu/~steinh/naturalquasicrystals.html. They include Luca Bindi and Paul J. Steinhardt, "The Discovery of the First Natural Quasicrystal: A New Era for Mineralogy?" *Elements* (newsletter), February 2012, doi:10.2113/gselements.8.1.13, http://www.physics.princeton.edu/~steinh/Bindi%20&%20Steinhardt_2012_ELEMENTS.pdf; Chi Ma, Chaney Lin, Luca Bindi, and Paul J. Steinhardt, "Hollisterite (Al3Fe), Kryachkoite (Al,Cu)6(Fe,Cu), and Stolperite (AlCu): Three New Minerals from the Khatyrka CV3 Carbonaceous Chondrite," *American Mineralogist* 102 (2017): 690–93.

30. "Khatyrka," Meteoritical Bulletin Database, https://www.lpi.usra.edu/meteor/metbull.php?code=55600. All minerals have to be certified as new by the Commission on New Minerals Nomenclature and Classification, a part of the International Mineralogical Association, which also approves new mineral names. Jeffrey de Fourestier, "The Naming of Mineral Species Approved by the Commission on New Minerals and Mineral Names of the International Mineralogical Association: A Brief History," *Canadian Mineralogist* 40, no. 6 (2002): 1721–35, doi:10.2113/gscanmin.40.6.1721.

31. The Z machine is "now able to propel small plates at 34 kilometers a second, faster than the 30 km/sec that Earth travels through space in its orbit about the sun." Mike Hanlon, "Sandia Lab's Z Machine: The Fastest Gun in the West," *New Atlas*, 10 June 2005, https://newatlas.com/go/4143/. It is hard to find these numbers on the official Z machine website: https://www.sandia.gov/z-machine/.

32. See the impressive list on Professor Ma's website: http://www.its.caltech.edu/~chima/.

33. Dennis V. Byrnes, James M. Longuski, and Buzz Aldrin, "Cycler Orbit between Earth and Mars," *Journal of Spacecraft and Rockets* 30, no. 3 (1993): 334.

34. Nathan Strange, Damon Landau, Paul Chodas, and James Longuski, "Identification of Retrievable Asteroids with the Tisserand Criterion," *Aerospace Research Central*, 1 August 2014, AIAA 2014-4458, https://doi.org/10.2514/6.2014-4458; Nathan Strange, Damon Landau, and James Longuski, *Redirection of Asteroids onto Earth-Mars Cyclers*, Advances in the Astronautical Sciences: Spaceflight Mechanics 2015, vol. 155, AAS 15-462, available at http://www.univelt.com/book=5109.

5. Love

1. "With approximately 24,000 attendees in 2016, AGU's Fall Meeting is the largest Earth and space science meeting in the world": AGU.org, https://fallmeeting.agu.org/2016/. Compare this with the 400 or so at "Letters of Intent for 2011," International Astronomical Union, 2011, https://www.iau.org/science/meetings/future/loi_2011/loi10/, under "Rationale."

2. Near the edge of the asteroid the eclipse will be shorter, so it takes an array of telescopes around the path of the eclipse to find the size. With a half dozen telescopes we can get an idea of the shape of the asteroid, not just its size.

3. *Catalina Sky Survey*: https://catalina.lpl.arizona.edu; *Pan-STARRS—The Panoramic Survey Telescope and Rapid Response System*: https://neo.ifa.hawaii.edu.

4. J. L. Galache, C. L. Beeson, K. K. McLeod, and M. Elvis, "The Need for Speed in Near-Earth Asteroid Characterization," *Planetary and Space Science* 111 (2015): 155.

5. Peter Vereš et al., "Unconfirmed Near-Earth Objects," *Astronomical Journal* 156 (2018): 5.

6. Benoit Carry, "Density of Asteroids," *Planetary and Space Science* 73 (2012): 98.

7. P. Bartczak and G. Dudziński, "Shaping Asteroid Models Using Genetic Evolution (SAGE)," *Monthly Notices of the Royal Astronomical Society* 473, no. 4 (2019): 5050–65.

8. There were 1,109 radar-detected asteroids and comets as of 9 June 2020: Radar-Detected Asteroids and Comets, https://echo.jpl.nasa.gov/asteroids/index.html.

9. "Asteroid and Comet Mission Targets Observed by Radar," last updated 3 September 2020, Jet Propulsion Laboratory, NASA, https://echo.jpl.nasa.gov/~lance/radar.small.body.mission.targets.html; "List of Missions to Minor Planets," Wikipedia, https://en.wikipedia.org/wiki/List_of_missions_to_minor_planets (accessed 4 September 2020).

10. NASA Galileo website: https://solarsystem.nasa.gov/missions/galileo/overview/.

11. ESA Asteroid (21) Lutetia: https://sci.esa.int/web/rosetta/-/47389-21-lutetia.

12. "NEAR Shoemaker," NASA, https://solarsystem.nasa.gov/missions/near-shoemaker/in-depth/.

13. H. C. Wilson, "Measuring the Distance of the Sun by Means of the Planet Eros," *Popular Astronomy* 12 (1904), http://articles.adsabs.harvard.edu/pdf/1904PA. . . . .12..149W.

14. "Dawn Mission Overview," NASA TV, https://www.nasa.gov/mission_pages/dawn/mission/index.html (accessed 4 September 2020).

15. "About Asteroid Explorer 'HAYABUSA' (MUSES-C)," Japan Aerospace Exploration Agency (JAXA), https://global.jaxa.jp/projects/sas/muses_c/ (accessed 4 September 2020).

16. "About Asteroid Explorer 'Hayabusa2,' " JAXA, https://global.jaxa.jp/projects/sas/hayabusa2/ (accessed 4 September 2020).

17. S. Watanabe et al., "Hayabusa2 Arrives at the Carbonaceous Asteroid 162173 Ryugu—A Spinning Top–Shaped Rubble Pile," *Science* 364, no. 6437 (2019): 268–72.

18. Planetary Society, "Nine-Year-Old Names Asteroid Target of NASA Mission in Competition Run by the Planetary Society" (2013), https://www.planetary.org/pressroom/releases/2013/nine-year-old-names-asteroid.html.

19. Mike Wehner, "NASA's Asteroid Probe Already Found Water on Bennu," *BGR* (10 December 2018), https://bgr.com/2018/12/10/bennu-water-asteroid-nasa-osiris-rex/; V. E. Hamilton et al., "Evidence for Widespread Hydrated Minerals on Asteroid (101955) Bennu," *Nature Astronomy* 3 (2019): 332–40.

20. Richard A. Lovett, "Asteroid Bennu Is Flinging Rocks into Space: OSIRIS-REx's Target Turns out to Be Very Rare, and Very Active, Posing Problems for the Mission," *Cosmos Magazine*, 19 March 2019, https://cosmosmagazine.com/space/asteroid-

bennu-is-flinging-rocks-into-space; Dante S. Lauretta et al., "Episodes of Particle Ejection from the Surface of the Active Asteroid (101955) Bennu," *Science* 366 (2019): 3544.

21. Dante S. Lauretta et al., "The Unexpected Surface of Asteroid (101955) Bennu," *Nature* 568 (2019): 55–60.

22. "Lucy: The First Mission to the Trojan Asteroids," NASA, https://www.nasa.gov/content/goddard/lucy-overview; H. F. Levison et al., "*Lucy*: Surveying the Diversity of the Trojan Asteroids: The Fossils of Planet Formation," *Lunar and Planetary Science* 48 (2017): 48.

23. "Psyche: Mission to a Metal World," Jet Propulsion Laboratory, https://www.jpl.nasa.gov/missions/psyche/; L. T. Elkins-Tanton and J. F. Bell III, "NASA's Discovery Mission to (16) Psyche: Visiting a Metal World," 2017, *European Planetary Science Congress* (EPSC), 11, 384, https://meetingorganizer.copernicus.org/EPSC2017/EPSC2017-384.pdf (accessed 4 September 2020).

24. John R. Brophy et al., *Asteroid Retrieval Feasibility Study*, Keck Institute for Space Studies (2012), California Institute of Technology, Jet Propulsion Laboratory, Pasadena, https://www.kiss.caltech.edu/final_reports/Asteroid_final_report.pdf.

25. Peter Jenniskens, Muawia H. Shaddad, et al., "The Impact and Recovery of Asteroid 2008 TC3," *Nature* 458 (2009): 485–88.

26. The Near Earth Objects Dynamic Site (NEODyS-2) lists them at these URLs: 2008TC3: https://newton.spacedys.com/neodys/index.php?pc=1.1.8&n=2008TC3; 2014AA: https://newton.spacedys.com/neodys/index.php?pc=1.1.8&n=2014AA; 2018LA: https://newton.spacedys.com/neodys/index.php?pc=1.1.8&n=2018LA; 2019MO: https://newton.spacedys.com/neodys/index.php?pc=1.1.8&n=2019MO.

27. Peter Jenniskens, "The 2002 Leonid MAC Airborne Mission: First Results," *WGN: The Journal of the International Meteor Organization* 30 (2002): 6.

6. Fear

1. Actually, what Sun Tzu said was much more subtle: "Hence the saying: If you know the enemy and know yourself, you need not fear the result of a hundred battles. If you know yourself but not the enemy, for every victory gained you will also suffer a defeat." *Art of War*, 18, https://www.suntzuonline.com/chapter-3-attack-by-stratagem/.

2. Alan W. Harris, "What Spaceguard Did," *Nature* 453 (2008): 1178.

3. JPL Center for NEO Studies, Discovery Statistics: https://cneos.jpl.nasa.gov/stats/site_all.html.

4. Minor Planet Center: https://minorplanetcenter.net.

5. Vera C. Rubin Observatory: https://www.aura-astronomy.org/centers/nsfs-oir-lab/rubinobservatory/; Legacy Survey of Space and Time (LSST): https://www.lsst.org; Thomas H. Zurbuchen, "Planetary Defense Strategy," NASA Science Mission Directorate, 23 September 2019, https://secureservercdn.net/198.71.233.197/b13.8cb.myftpupload.com/wp-content/uploads/2019/09/Thomas-Z-planetary-defense-strategy-Sep-23-2019.pdf; Marcia Smith, "NASA Announces New Mission to Search

for Asteroids," Space Policy Online, 23 September 2019, https://spacepolicyonline. com/news/nasa-announces-new-mission-to-search-for-asteroids/. The mission was previously called NEOCAM: https://neocam.ipac.caltech.edu/page/mission.

6. Amy Mainzer et al., "NEOWISE Observations of Near-Earth Objects: Preliminary Results," *Astrophysical Journal* 743 (2011): 156.

7. NASA Wide-field Infrared Explorer Mission: https://www.nasa.gov/mission_pages/WISE/main/index.html; Edward L. Wright et al., "The Wide-Field Infrared Survey Explorer (WISE): Mission Description and Initial On-Orbit Performance," *Astronomical Journal* 140, no. 6 (2010): 1868–81.

8. ESA Gaia mission: https://sci.esa.int/web/gaia/-/28820-summary; Gaia Collaboration, "The Gaia Mission," *Astronomy & Astrophysics* 595 (2016): 1.

9. Željko Ivezić et al., "Solar System Objects Observed in the Sloan Digital Sky Survey Commissioning Data," *Astrophysical Journal* 122 (2001): 2749.

10. A. Thirouin, N. Moskovitz, et al., "The Mission Accessible Near-Earth Objects Survey (MANOS): First Photometric Results," *Astronomical Journal* 152, no. 6 (2016): 163.

11. Or so it is often said. Expert opinion says that hitting sharks on their pain-sensitive gills or eyes works best: "How to Fend off Sharks," *How to Guides 365,* https://www.howtoguides365.com/how-to/fend-off-sharks/ (accessed 4 September 2020).

12. NASA Double Asteroid Redirection Test (DART) mission: https://www.nasa.gov/planetarydefense/dart; Double Asteroid Redirection Test—NASA's First Planetary Defense Mission: https://dart.jhuapl.edu; Mike Wall, "ESA *Hera* Mission: Europe Officially Signs on for Asteroid-Smashing Effort," Space.com, 3 December 2019, https://www.space.com/european-hera-asteroid-mission-approved.html; Ahmed Bilal, "ESA AIDA Mission: The ESA Cancels AIM to Gather Half a Billion Dollars for the Next Mars Lander Despite Recent Crash," wccftech, 6 December 2016, https://wccftech.com/esa-gives-half-billion-exomars-project/.

13. Lutz D. Schmadel, *Dictionary of Minor Planet Names* (Heidelberg: Springer-Verlag, 2006).

14. Maria Temming, "An Asteroid's Moon Got a Name So NASA Can Bump It off Its Course," *Science News,* 30 June 2020, https://www.sciencenews.org/article/asteroid-moon-name-nasa-course-deflection-mission.

15. "Facts and Figures," Burj Khalifa, https://www.burjkhalifa.ae/en/the-tower/facts-figures// (accessed 4 September 2020).

16. Angie Yee, "Speed of a Snail," in *The Physics Fact Book,* https://hypertextbook.com/facts/1999/AngieYee.shtml (accessed 4 September 2020).

17. Cathy Plesko, "Stopping an Earth-Bound Asteroid in Its Tracks," Space.com, 18 April 2019, https://www.space.com/stopping-earth-bound-asteroid-op-ed.html.

18. Edward T. Lu and Stanley G. Love, "Gravitational Tractor for Towing Asteroids," *Nature* 438 (2005): 177.

19. Rachel Shweky, "Speed of a Turtle or a Tortoise," in *The Physics Fact Book,* https://hypertextbook.com/facts/1999/RachelShweky.shtml (accessed 4 September 2020).

20. Daniel D. Mazanek et al., "Enhanced Gravity Tractor Technique for Planetary Defense," 4th IAA Planetary Defense Conference—PDC 2015 (13–17 April 2015), Frascati, Rome, IAA-PDC-15-04-11: https://ntrs.nasa.gov/archive/nasa/casi.ntrs.nasa.gov/20150010968.pdf.

21. Mike Wall, "Asteroid Billiards: This Wild Idea to Protect Earth Just Might Work," Space.com, 24 August 2018, https://www.space.com/41592-asteroid-billiards-smashing-dangerous-space-rocks.html.

7. Greed

1. M. J. Sonter, "The Technical and Economic Feasibility of Mining the Near-Earth Asteroids," *Acta Astronautica* 41 (1997): 637–47.

2. Ontario Ministry of Energy, Northern Development and Mines, *Exploration and Developing Minerals in Ontario,* last modified 10 May 2019, https://www.mndm.gov.on.ca/en/mines-and-minerals/exploration-and-developing-minerals-ontario.

3. Steve Squyres, *Roving Mars: Spirit, Opportunity, and the Exploration of the Red Planet* (New York: Hyperion, 2005).

4. Alan Boyle, "Gradatim Ferociter! Jeff Bezos Explains Blue Origin's Motto, Logo . . . and the Boots," Geekwire, 24 October 2016, https://www.geekwire.com/2016/jeff-bezos-blue-origin-motto-logo-boots/.

5. See Colorado Division of Reclamation, Mining and Safety, "AUGER," https://gis.colorado.gov/dnrviewer/Index.html?viewer=drms (accessed 1 October 2020).

6. Douglas A. Vakoch and Matthew F. Dowd, eds., *The Drake Equation: Estimating the Prevalence of Extraterrestrial Life through the Ages* (Cambridge: Cambridge University Press, 2015).

7. Martin Elvis, "How Many Ore-Bearing Asteroids?" *Planetary and Space Sciences* 91 (2014): 20, https://arxiv.org/pdf/1312.4450.pdf.

8. Jacob Aron, "Alien-Hunting Equation Revamped for Mining Asteroids," *New Scientist,* 4 December 2013, https://www.newscientist.com/article/dn24696-alien-hunting-equation-revamped-for-mining-asteroids/#ixzz6Pih15zu; " 'Elvis Equation' Estimates Number of Asteroids Worth Mining (Spoiler: Not Very Many)," *The Physics arXiv Blog,* medium.com, 8 January 2014, https://medium.com/the-physics-arxiv-blog/elvis-equation-estimates-number-of-asteroids-worth-mining-spoiler-not-very-many-e0063699d199.

9. Joseph Scott Stuart and Richard P. Binzel, "Bias-Corrected Population, Size Distribution, and Impact Hazard for the Near-Earth Objects," *Icarus* 170 (2004): 295–311.

10. Eugene Jarosewich, "Chemical Analyses of Meteorites: A Complication of Stony and Iron Meteorite Analyses," *Meteoritics* 25 (1990): 323–37.

11. "The Number of Asteroids We Could Visit and Explore Has Just Doubled," *Universe Today,* 11 February 2015, https://www.universetoday.com/tag/nasa-neo/; "Accessible NEAs," NASA Center for Near-Earth Object Studies, https://cneos.jpl.nasa.gov/nhats/ (accessed 4 September 2020).

12. Calculations by Lance Benner in "Near-Earth Asteroid Delta-V for Spacecraft Rendezvous," NASA, https://cneos.jpl.nasa.gov/nhats/ (accessed 5 September 2020).

13. Anthony Taylor, Jonathan C. McDowell, and Martin Elvis, "A Delta-V Map of the Known Main Belt Asteroids," *Acta Astronautica* 146 (2018): 73–82.

14. Richard Schodde, "Trends in Exploration," MinEx Consulting (presentation at the Imarc Conference, Melbourne, October 2019), http://minexconsulting.com/trends-in-exploration/.

15. Sabrina Cooper, "Remember When Linda Evangelista Made Waking Up a Five-Figure Act? In Honor of the Supermodel's 54th Birthday This Week," *CR Fashion Book*, May 9, 2019.

16. Amy Mainzer et al., "NEOWISE Observations of Near-Earth Objects: Preliminary Results," *Astrophysical Journal* 743 (2011): 156.

17. Francesca E. DeMeo et al., "An Extension of the Bus Asteroid Taxonomy into the Near-Infrared," *Icarus* 202 (2009): 160.

18. "ATLAS: The Asteroid Terrestrial-Impact Last Alert System," Institute for Astronomy, University of Hawaii, 15 February 2013, http://www.ifa.hawaii.edu/info/press-releases/ATLAS/; John Tonry et al. "ATLAS: A High-Cadence All-Sky Survey System," *Publications of the Astronomical Society of the Pacific* 130, no. 988 (2018): 064505.

19. Peter Jenniskens et al., "Radar-Enabled Recovery of the Sutter's Mill Meteorite, a Carbonaceous Chondrite Regolith Breccia," *Science* 338 (2012): 1583.

20. Marshall Trimble, "What's the Story behind the Phrase 'There's Gold in Them Thar Hills'?" *True West*, 18 August 2015, https://truewestmagazine.com/whats-the-story-behind-the-phrase-theres-gold-in-them-thar-hills/. Trimble, the Arizona official historian, states this saying is from Mark Twain's 1892 novel *The American Claimant*.

21. Writing was invented to keep track of accounts. See, for example, Ira Spar, "The Origins of Writing," Heilbrunn Timeline of Art History, Metropolitan Museum of Art, October 2004, https://www.metmuseum.org/toah/hd/wrtg/hd_wrtg.htm.

22. Michael Shao et al., "Finding Very Small Near-Earth Asteroids Using Synthetic Tracking," *Astrophysical Journal* 782 (2014): 1; B612 Foundation: https://b612foundation.org.

23. Jonathan C. McDowell, "The Low Earth Orbit Satellite Population and Impacts of the SpaceX Starlink Constellation," *Astrophysical Journal* 892 (2020): L36.

24. Colorado School of Mines, Golden, Colorado, Space Resources Program: https://space.mines.edu; Luleå University of Technology, Kiruna, Sweden, Onboard Space Systems: https://www.ltu.se/research/subjects/Rymdtekniska-system?l=en.

25. Regolith X-ray Imaging Spectrometer (REXIS): https://www.asteroidmission.org/?attachment_id=1205#main; R. A. Masterson et al., "Regolith X-Ray Imaging Spectrometer (REXIS) aboard the OSIRIS-REx Asteroid Sample Return Mission," *Space Science Reviews* 214, no. 1 (2018), https://doi.org/10.1007/s11214-018-0483-8.

26. OSIRIS-REx Laser Altimeter (OLA): https://www.asteroidmission.org/?attachment_id=1201#main.

27. Luke Dormehl, "Meet the Tech That Revealed a Hidden Chamber inside Egypt's Great Pyramid," DigitalTrends, 4 November 2017. Wikipedia gives a good, more technical, overview: "Muon Tomography," https://en.wikipedia.org/wiki/Muon_tomography (accessed 5 September 2020).

28. Bob Yirka, "Fix for Mars Lander 'Mole' May Be Working," Phys.org, 8 June 2020, https://phys.org/news/2020-06-mars-lander-mole.html. For information on the Heat Flow and Physical Properties Package instrument on the *InSight*, see "Instruments," NASA Mars InSight Mission, https://mars.nasa.gov/insight/spacecraft/instruments/hp3/ (accessed 5 September 2020).

29. Mike Wall, "Asteroid Miners' Arkyd-6 Satellite Aces Big Test in Space," Space.com, 25 April 2018, https://www.space.com/40400-planetary-resources-asteroid-mining-satellite-mission-accomplished.html.

30. "What is the Deep Space Network?" NASA TV, 30 March 2020, https://deepspace.jpl.nasa.gov/about/. You can see what DSN is observing right now at "Deep Space Network Now," Jet Propulsion Laboratory, NASA, https://eyes.nasa.gov/dsn/dsn.html (accessed 5 September 2020).

31. "NASA TV—Ion Propulsion," NASA TV Space Tech, 11 January 2016, https://www.nasa.gov/centers/glenn/about/fs21grc.html.

32. David Hamen, "What Makes Time on the DSN So Expensive?" Space Exploration Stack Exchange, https://space.stackexchange.com/questions/21005/what-makes-time-on-the-dsn-so-expensive (accessed 4 September 2020). The calculation is based on "NASA's Mission Operations and Communications Services," NASA, 1 October 2014, https://deepspace.jpl.nasa.gov/files/6_NASA_MOCS_2014_10_01_14.pdf.

33. P. Graven et al., "XNAV for Deep Space Navigation," in *31st Annual AAS Guidance and Control Conference,* American Astronautical Society, 1–6 February 2008, http://www.asterlabs.com/publications/2008/Graven_et_al,_AAS_31_GCC_February_2008.pdf.

34. "NASA Team First to Demonstrate X-Ray Navigation in Space," NASA, 11 January 2018, https://www.nasa.gov/feature/goddard/2018/nasa-team-first-to-demonstrate-x-ray-navigation-in-space; Jason W. Mitchell, "SEXTANT X-Ray Pulsar Navigation Demonstration: Initial On-Orbit Results," 41st Annual American Astronautical Society (AAS) Guidance and Control Conference, 1 February 2018, https://ntrs.nasa.gov/archive/nasa/casi.ntrs.nasa.gov/20180001252.pdf.

35. Robert Zimmerman, "Docking in Space," *Invention and Technology* 17, no. 2 (2001), https://www.inventionandtech.com/content/docking-space-1.

36. Richard B. Setlow, "The Hazards of Space Travel," Science and Society: Viewpoint, *European Molecular Biology Organization Report,* 4, no. 11 (November 2003): 1013–16, doi:10.1038/sj.embor.7400016.

37. NASA, "Space Shuttle Orbital Docking System," Math and Science @ work, AP* Physics Educator Edition, https://www.nasa.gov/pdf/593864main_AP_ED_Phys_ShuttleODS.pdf.

38. "Percentage of Total Population Living in Coastal Areas," United Nations, https://www.un.org/esa/sustdev/natlinfo/indicators/methodology_sheets/oceans_seas_coasts/pop_coastal_areas.pdf (accessed 5 September 2020).

39. *NASA's Recommendations to Space-Faring Entities: How to Protect and Preserve the Historic and Scientific Value of U.S. Government Lunar Artifacts,* 20 July 2011, http://www.collectspace.com/news/NASA-USG_lunar_historic_sites.pdf.

40. D. Hestroffer et al., "Small Solar System Bodies as Granular Media," *Astronomy and Astrophysics Review* 27, no. 1 (2019), https://rdcu.be/bHM9E; doi:10.1007/s00159-019-0117-5 (accessed 5 September 2020).

41. "Granular Materials," *Complex Systems*, Physics Department, University of California, Santa Barbara, http://web.physics.ucsb.edu/~complex/research/granular.html (accessed 5 September 2020).

42. Takeo Watanabe et al., "Penetration Dynamics of an Asteroid Sampling System Inspired by Japanese Sword Technology," *Transactions of the Japan Society for Aeronautical and Space Sciences* 14, no. 30 (2016): Pk_23–Pk_28.

43. John S. Lewis, *Mining the Sky: Untold Riches from the Asteroids, Comets, and Planets* (Reading, MA: Addison-Wesley, 1996).

44. "The electrical power required to operate the *Chandra* spacecraft and instruments is 2 kilowatts, about the same power as a hair dryer." "Top 10 Facts about Chandra," Chandra X-Ray Science Center, https://chandra.harvard.edu/about/top_ten.html (accessed 5 September 2020).

45. Nirlipta P. Nayak and Bhatu K. Pal, "Separation Behaviour of Iron Ore Fines in Kelsey Centrifugal Jig," *Journal of Minerals and Materials Characterization and Engineering* 1 (2013): 85–89, http://dx.doi.org/10.4236/jmmce.2013.13016.

46. Frank K. Crundwell, Michael Moats, Venkoba Ramachandran, Timothy Robinson, and W. G. Davenport, *Extractive Metallurgy of Nickel, Cobalt and Platinum Group Metals* (Amsterdam: Elsevier, 2011).

47. Emily Flashman, "How Plastic-Eating Bacteria Actually Work—A Chemist Explains," The Conversation, 19 April 2018, https://phys.org/news/2018-04-plastic-eating-bacteria-worka-chemist.html; Shosuke Yoshida, Kazumi Hiraga, Toshihiko Takehana, Ikuo Taniguchi, Hironao Yamaji, Yasuhito Maeda, Kiyotsuna Toyohara, Kenji Miyamoto, Yoshiharu Kimura, and Kohei Oda, "A Bacterium That Degrades and Assimilates Poly(ethylene Terephthalate)," *Science* 351, no. 6278 (2016): 1196–99, doi:10.1126/science.aad6359.

48. Michael Klas, N. Tsafnat, J. Dennerley, S. Beckman, B. Osborne, A. G. Dempster, and M. Manefield, "Biomining and Methanogenesis for Resource Extraction from Asteroids," *Space Policy* 34 (2015): 18–22.

49. R. Volger, G. M. Pettersson, S. J. J. Brouns, L. J. Rothschild, A. Cowley, and B. A. E. Lehner, "Mining Moon & Mars with Microbes: Biological Approaches to Extract Iron from Lunar and Martian Regolith," *Planetary and Space Science* 184 (2020): 104850.

50. *Gridded Ion Thrusters (NEXT-C)*, NASA Glenn Research Center: https://www1.grc.nasa.gov/space/sep/gridded-ion-thrusters-next-c/.

51. Momentus Space: https://momentus.space; press releases: https://momentus.space/press/.

52. Marc Montgomery, "Soviet Radiation across the Arctic," *Canada History*, Radio Canada International, 24 January 1978, last updated 29 January 2017, https://www.rcinet.ca/en/2017/01/24/canada-history-jan-24-1978-soviet-radiation-across-the-arctic/.

53. Marc Gibson, David Poston, Patrick McClure, Thomas Godfroy, Maxwell Briggs, and James Sanzi, "The Kilopower Reactor Using Stirling Technology (KRUSTY) Nuclear Ground Test Results and Lessons Learned," *NASA Technical Report* (2018), https://ntrs.nasa.gov/archive/nasa/casi.ntrs.nasa.gov/20180005435.pdf.

54. Stephanie Thomas, "Fusion Drive for Rapid Deep Space Propulsion," Future In-Space Operations (FISO), 1 June 2019, http://www.psatellite.com/tag/dfd/; Stephanie J. Thomas, Michael Paluszek, Charles Swanson, Samuel Cohen, and Slava G. Turyshev, "Fusion Propulsion and Power for Extrasolar Exploration," *70th International Astronautical Congress (IAC), 2019*, Symposium C3, session, 5-C4.7, paper 10, https://collaborate.princeton.edu/en/publications/fusion-propulsion-and-power-for-extrasolar-exploration.

55. Small Explorer missions had a cost cap of $145 million in 2019. *Announcement of Opportunity Astrophysics Explorers Program 2019 Small Explorer (SMEX)*, Full Missions Evaluation Plan, NASA, 16 April 2019, https://explorers.larc.nasa.gov/2019APSMEX/SMEX/pdf_files/2019-Astro-SMEX%20Eval-Plan-rev3_FINAL.pdf.

56. "Lunar Gravity Assists for Asteroids," Permanent.com, https://www.permanent.com/space-transportation-lunar-gravity-assist.html (accessed 5 September 2020); A. Ledkov, N. Eismont, R. Nazirov, and M. Boyarsky, "Near Moon Gravity Assist Maneuvers as a Tool for Asteroid Capture onto Earth Satellite Orbit," International Symposium on Space Flight Dynamics (ISSFD), 2015, https://issfd.org/2015/files/downloads/papers/073_Ledkov.pdf.

57. Radu Dan Rugescu, "Tether Balute Low Cost De-Orbit and Recovery Project," *AIAA 2nd Responsive Space Conference*, Los Angeles, April 19–22, 2004, https://www.researchgate.net/publication/228382475_TETHER_BALUTE_LOW_COST_DE-ORBIT_AND_RECOVERY_PROJECT.

58. We were lucky that the main engines of the Space Shuttle landed in a swamp; otherwise people could have been hurt. "Cause and Consequences of the Columbia Disaster," *Space Safety Magazine*, 6 May 2014, http://www.spacesafetymagazine.com/space-disasters/columbia-disaster/columbia-tragedy-repeated/; NASA, *Report of Columbia Accident Investigation Board* (CAIB), 2003, https://www.nasa.gov/columbia/home/CAIB_Vol1.html.

8. Getting Space off the Ground

1. Bryce Space and Technology, *2018 Global Space Economy Report*, https://brycetech.com/reports.

2. Using the cost for 20 hours' flying at $10k/hour for 12 passengers quoted at Singapore Private Jet Charter, Paramount Business Jets: https://www.paramountbusinessjets.com/cities/singapore-singapore.html (accessed 20 May 2020).

3. Remarks by Elon Musk at the National Press Club, The Future of Human Spaceflight, Washington, DC, 29 September 2011, archived video (original not available) at https://www.c-span.org/video/?301817-1/future-human-space-flight.

4. The *Saturn V* could put 118,000 kilograms into LEO at a cost of $1.2 billion (2016 dollars) per launch, and so at a cost of $10,000 per kilogram. *Falcon Heavy* launches 63,800 kilograms at a cost of $90 million, so $1,500 per kilogram. See Matt Williams, "Falcon Heavy vs. Saturn V," *Universe Today*, 25 July 2016, https://www.universetoday.com/129989/saturn-v-vs-falcon-heavy/.

5. The Space Shuttle cost, including all development expenditures, came to $1.6 billion per launch to get up to 29,000 kilograms to LEO, implying a cost of $55,000 per kilogram. In the later years the incremental cost per launch was $750 million, or $26,000 per kilogram. See Mike Wall, "NASA's Shuttle Program Cost $209 Billion—Was It Worth It?" Space.com, 5 July 2011, https://www.space.com/12166-space-shuttle-program-cost-promises-209-billion.html; Leonard David, "Space Shuttle Cost Gets a Reality Check," NBCNEWS.com, 11 February 2005, http://www.nbcnews.com/id/6953606/ns/technology_and_science-space/t/space-shuttle-cost-gets-reality-check/#.Xu5rCy-ZPYI; Roberto Galvez, Stephen Gaylor, Charles Young, Nancy Patrick, Dexer Johnson, and Jose Ruiz, "The Space Shuttle," *The Space Shuttle and Its Operations*, NASA, https://www.nasa.gov/centers/johnson/pdf/536822main_Wings-ch3.pdf (accessed 5 September 2020).

6. SpaceX, Capabilities and Services, https://www.spacex.com/media/Capabilities&Services.pdf (accessed 15 July 2020).

7. Michael Sheetz, "NASA's Deal to Fly Astronauts with Boeing Is Turning out to Be Much More Expensive Than SpaceX," *CNBC*, 19 November 2019, https://www.cnbc.com/2019/11/19/nasa-cost-to-fly-astronauts-with-spacex-boeing-and-russian-soyuz.html; Stephen Clark, "NASA Inks Deal with Roscosmos to Ensure Continuous U.S. Presence on Space Station," Spaceflight Now, 12 May 2020, https://spaceflight-now.com/2020/05/12/nasa-inks-deal-with-roscosmos-to-ensure-continuous-u-s-presence-on-space-station/.

8. "Movie Budget and Financial Performance Records," The Numbers, https://www.the-numbers.com/movie/budgets (accessed 5 September 2020); Brent Lang and Justin Kroll, "Leonardo DiCaprio, Jennifer Lawrence and Other Star Salaries Revealed," *Variety*, 8 May 2018, https://variety.com/2018/film/news/celebrity-salaries-daniel-craig-jennifer-lawrence-leonardo-dicaprio-1202801717/; Eli Glasner, "Mission Possible: How Tom Cruise's Plan to Film in Space Fits NASA's Trajectory," *CBC News*, 30 May 2020, https://www.cbc.ca/news/entertainment/tom-cruise-nasa-mission-1.5590982.

9. William Harwood, "SpaceX and Space Adventures to Launch Space Tourism Flight in 2022," *CBS News*, 18 February 2020, https://www.cbsnews.com/news/spacex-space-adventures-tourism-orbit-crew-dragon-2022/. Are you interested in a ticket? SpaceX says, "For inquiries about our private passenger program, contact sales@spacex.com" (https://www.spacex.com/human-spaceflight/.) Or, in a sign of how routine this could become, fill in the "Contact Us" form at SpaceAdventures: https://spaceadventures.com/experiences/low_earth_orbit/. Please don't bother them unless you are willing and able to drop a few million dollars on the ride.

10. Stephen Clark, "Axiom Strikes Deal with SpaceX to Ferry Private Astronauts to Space Station," Spaceflight Now, 5 March 2020, https://spaceflightnow.

com/2020/03/05/axiom-strikes-deal-with-spacex-to-ferry-private-astronauts-to-space-station/. For details, see Private Astronaut Missions on the Axiom website: https://www.axiomspace.com/private-astronauts-missions.

11. Michael Sheetz, "Virgin Galactic Is Seeing Strong Demand for Tourist Flights to Space, Will Re-open Ticket Sales," *CNBC* (9 January 2020), https://www.cnbc.com/2020/01/09/virgin-galactic-ticket-sales-will-re-open-this-year-ceo-says.html.

12. *Booster Landing Falcon Heavy Feb 6, 2018*, YouTube, https://www.youtube.com/watch?v=pX6oGB3nv1I.

13. Alan Boyle, "Billionaire-Backed Asteroid Mining Venture Starts with Space Telescopes," *NBC News*, 23 April 2012, https://www.nbcnews.com/science/cosmic-log/billionaire-backed-asteroid-mining-venture-starts-space-telescopes-flna731384.

14. *Blue Origin's New Shepard First Landing*, YouTube, November 24, 2015, https://www.youtube.com/watch?v=sij4ivRwHuQ.

15. Eric Ralph, "SpaceX Rings in Falcon 9's 10th Anniversary with a Rocket Reusability First," Teslarati.com, 5 June 2020, https://www.teslarati.com/spacex-falcon-9-10th-anniversary-rocket-record/. This is something of a fan site but appears to be factually correct.

16. Mike Wall, "SpaceX's Starship May Fly for Just $2 Million per Mission, Elon Musk Says," Space.com, 6 November 2019, https://www.space.com/spacex-starship-flight-passenger-cost-elon-musk.html. If Musk's claim is correct, *Starship* can carry 100 people, so the cost to SpaceX is just $20,000 each. That's not going to be the ticket price, but even if SpaceX put a five times markup on the launch cost, charging $10 million/launch, a ticket would still cost only $100,000.

17. Flying was about 20 times more dangerous in the 1970s: "The five-year moving average fatality rate has improved by an astonishing 96% between the mid 1970s and the late 2000s." I. Savage, "Comparing the Fatality Risks in United States Transportation across Modes and over Time," *Research in Transportation Economics* 43 (2013): 9–22, https://faculty.wcas.northwestern.edu/~ipsavage/436.pdf.

18. Blue Origin's first orbital-class rocket is *New Glenn*, "featuring a reusable first stage built for 25 missions," "able to launch and land in 95% of weather conditions, making it a reliable option for payload customers": "New Glenn," *Blue Origin*, https://www.blueorigin.com/new-glenn/ (accessed 5 September 2020). ULA's new rocket is the *Vulcan Centaur*. It does not (yet) incorporate reuse, but ULA claims that its "simple design is more cost-efficient for all customers." "Vulcan Centaur," ULA, https://www.ulalaunch.com/rockets/vulcan-centaur (accessed 5 September 2020); Arianespace is building the new *Ariane 6*, which is not initially reusable but was "conceived for reduced production costs and design-to-build lead times": Arianespace, https://www.arianespace.com/ariane-6/ (accessed 5 September 2020).

19. Roger D. Launius and Howard E. McCurdy, eds., *Seeds of Discovery: Chapters in the Economic History of Innovation within NASA* (2015), https://www.nasa.gov/sites/default/files/atoms/files/seeds_of_discovery_ms-spaceportal.pdf.

20. A detailed history of the COTS program (from a NASA perspective) is given in the anonymous *Commercial Orbital Transportation Services: A New Era in Spaceflight*,

NASA SP-2014-617, February 2014: https://www.nasa.gov/sites/default/files/files/
SP-2014-617.pdf.

21. Ashlee Vance, *Elon Musk: Tesla, SpaceX, and the Quest for a Fantastic Future*
(New York: Harper Collins, 2015), 252.

22. "The NASA Space Act Agreement: Partnering with NASA," *Commercial Technology Partnerships*, Jet Propulsion Laboratory, NASA, https://nsta.jpl.nasa.gov/com
mercial/saa.php (accessed 5 September 2020).

23. Mike Wall, "Private Orbital Sciences Rocket Explodes during Launch, NASA
Cargo Lost," Space.com, 28 October 2014, https://www.space.com/27576-private-
orbital-sciences-rocket-explosion.html.

24. Darrell Etherington, "SpaceX's CRS-7 Mission Ends in Catastrophic Failure, Loss
of Vehicle," Tech Crunch, 28 June 2015, https://techcrunch.com/2015/06/28/watch-
spacex-launch-crs-7-and-attempt-rocket-recovery-via-drone-live-now/?guccounter=
1&guce_referrer=aHRocHM6Ly93d3cuZ29vZ2xlLmNvbS8&guce_referrer_
sig=AQAAAIa3ny6xIQ9kI_ftPDMyQ19n7Nu-j-HRWyKqL95jE46HSQf2VfKLnlsuO7k-
ScnvYY-GpXXzYFcl8O3idA1ShqENpOCBOxoMXk1TVnGlCPfo1NhC1E3Yq-
tUovKjsnS7sXppaGZqB6MgmlND-fxj4RBtZIsHrH3UN9lC3ujcwFavNO.

25. Sarah Lewin, "Cygnus Spaceship Launch Restarts Orbital ATK Cargo Missions
for NASA," Space.com, 6 December 2015, https://www.space.com/31278-cygnus-
spacecraft-launch-orbital-atk-return-to-flight.html; "Photos: SpaceX Falcon 9 Rocket
Launch and Landing for CRS-8 Mission," Space.com, 10 April 2016, https://www.
space.com/32514-spacex-rocket-launch-landing-photos-dragon-crs8-mission.html.

26. Sierra Nevada Corporation, "NASA Selects Sierra Nevada Corporation's Dream
Chaser® Spacecraft for Commercial Resupply Services 2 Contract," press release, 14
January 2016, https://www.sncorp.com/press-releases/snc-crs2-announcement/.

27. Amanda Kooser and Stephen Shankland, "SpaceX's Historic Demo-2 Delivers
NASA Astronauts to ISS," c|net, 30 May 2020, https://www.cnet.com/how-to/
spacexs-historic-demo-2-delivers-nasa-astronauts-to-iss/.

28. Mike Wall, "NASA Teams with SpaceX, Blue Origin and More to Boost Moon
Exploration Tech," Space.com, 31 July 2019, https://www.space.com/nasa-moon-
mars-technology-commercial-partnerships.html; "NASA Names Companies to Develop
Human Landers for Artemis Moon Missions," NASA press release, 30 April 2020,
https://www.nasa.gov/press-release/nasa-names-companies-to-develop-human-
landers-for-artemis-moon-missions.

29. Jonathan C. McDowell, "The Edge of Space: Revisiting the Karman Line," *Acta
Astronautica* 151 (2018): 668–77, https://arxiv.org/abs/1807.07894.

30. Leonard David, "Inside ULA's Plan to Have 1,000 People Working in Space by
2045," Space Insider, 29 June 2016, https://www.space.com/33297-satellite-refueling-
business-proposal-ula.html.

31. Paul D. Spudis et al., "Initial Results for the North Pole of the Moon from Mini-
SAR, Chandrayaan-1 Mission," *Geophysical Research Letters* 37, no. 6 (2010): L06204.

32. Sandra Erwin, "In-Orbit Services Poised to Become Big Business," Space News,
10 June 2018, https://spacenews.com/in-orbit-services-poised-to-become-big-business/.

33. "Astroscale Brings Total Capital Raised to $191 Million, Closing Series E Funding Round," press release, 13 October 2020, Astroscale, https://astroscale.com/astroscale-brings-total-capital-raised-to-u-s-191-million-closing-series-e-funding-round/.

34. Honeybee Robotics, "The World Is Not Enough Demonstrates the Future of Space Exploration," press release, 15 January 2019, https://honeybeerobotics.com/wine-the-world-is-not-enough/?doing_wp_cron=1592747376.7651550769805908203125.

35. Caleb Henry, "Solar Panel Suppliers Adjust to GEO Satellite Slowdown," Space News, 24 January 2018, https://spacenews.com/solar-panel-suppliers-adjust-to-geo-satellite-slowdown/.

36. Bigelow Aerospace: http://bigelowaerospace.com; Leonard David, "Bigelow Aerospace's Genesis-1 Performing Well," Space.com, 21 July 2006, https://www.space.com/2649-bigelow-aerospace-genesis-1-performing.html; *Bigelow Expandable Activity Module*, Space Station Research Explorer, on NASA, https://www.nasa.gov/mission_pages/station/research/experiments/explorer/Investigation.html?#id=1579 (accessed 5 September 2020); Jeff Foust, "NASA Planning to Keep BEAM Module on ISS for the Long Haul," Space News, 12 August 2019, https://spacenews.com/nasa-planning-to-keep-beam-module-on-iss-for-the-long-haul/.

37. Lee Billings, "Who Will Build the World's First Commercial Space Station?" *Scientific American*, 26 May 2017, https://www.scientificamerican.com/article/who-will-build-the-world-rsquo-s-first-commercial-space-station/.

38. Axiom Space: https://www.axiomspace.com; for CEO Suffredini's experience, see "Axiom Space Overview and Team: The World's Leading Commercial Space Station Company," Axiom Space, https://www.axiomspace.com/overview-and-team (accessed 5 September 2020).

39. Michael Polanyi, *The Tacit Dimension* (New York: Anchor Books, 1967). See also "Michael Polanyi and Tacit Knowledge," Infed.org: Education, Community-Building and Change, http://infed.org/mobi/michael-polanyi-and-tacit-knowledge/ (accessed 5 September 2020).

40. Rich Smith, "Will Abandoning the International Space Station Set Investors Adrift?" Motley Fool, 24 February 2018, https://www.fool.com/investing/2018/02/24/will-abandoning-the-international-space-station-se.aspx.

41. Mike Wall, "1st Private Space Station Will Become an Off-Earth Manufacturing Hub," Space.com, 5 June 2017, https://www.space.com/37079-axiom-commercial-space-station-manufacturing.html.

42. Jeff Foust, "NASA Selects Axiom Space to Build Commercial Space Station Module," Space News, 28 January 2020, https://spacenews.com/nasa-selects-axiom-space-to-build-commercial-space-station-module/.

43. "Inside Nasa's New 'Space Home': Why the Philippe Starck–Designed Modules Resemble a 'Fetal Universe,' " Business Insider, 13 March 2020, https://www.scmp.com/magazines/style/news-trends/article/3074926/inside-nasas-new-space-home-why-philippe-starck; Natashah Hitti, "Philippe Starck Designs 'Foetal' Interiors for Axiom's Commercial Space Station," de zeen, 14 June 2018, https://www.dezeen.

com/2018/06/14/philippe-starck-designs-foetal-interiors-for-axioms-commercial-space-station/.

44. Jeff Foust, "Study Validates NanoRacks Concept for Commercial Space Station Module," Space News, 8 December 2017, https://www.space.com/39024-study-validates-nanoracks-concept-for-commercial-space-station-module.html.

45. Loren Grush, "How One Company Wants to Recycle Used Rockets into Deep-Space Habitats: Why Destroy When You Can Reuse?" The Verge, 14 June 2017, https://www.theverge.com/2017/6/14/15783494/nasa-nanoracks-ixion-nextstep-habitats-rocket-upper-stage.

46. Virginia P. Dawson and Mark D. Bowles, *Taming Liquid Hydrogen: The Centaur Upper Stage Rocket 1958–2002*, NASA History Series (2004): NASA SP-2004-4230, https://history.nasa.gov/SP-4230.pdf. The thickness of the *Centaur* tank walls is hard to find. The only place I could find it was in a tweet by the CEO of ULA: Tory Bruno, "The Amazing Centaur: America's Space Workhorse," Twitter, 23 May 2019, https://twitter.com/torybruno/status/1131638302761578496/photo/1.

47. Jonathan C. McDowell, "Atlas V Launch Table, JSR Launch Vehicle Database," Jonathan's Space Report, 14 June 2020, https://planet4589.org/space/lvdb/launch/Atlas5.

48. Jason Davis, "A Company You've Never Heard of Plans to Build the World's First Private Space Station," Planetary Society, 3 January 2017, http://www.planetary.org/blogs/jason-davis/2016/20170103-axiom-profile.html.

49. Sidney Perkowitz, "Bose-Einstein Condensate," *Encyclopedia Britannica*, https://www.britannica.com/science/Bose-Einstein-condensate (accessed 15 July 2020).

50. Dennis Becker et al., "Space-Borne Bose-Einstein Condensation for Precision Interferometry," *Nature* 562 (2018): 391–95, https://www.nature.com/articles/s41586-018-0605-1.

51. JPL Cold Atom Lab, "The Coolest Spot in the Universe," https://coldatomlab.jpl.nasa.gov; "Cold Atom Lab Creates Bose-Einstein Condensate on the ISS," NASA press release, 14 June 2020, http://spaceref.com/international-space-station/cold-atom-lab-creates-bose-einstein-condensate-on-the-iss.html.

52. V. I. Strelov et al., "Crystallization in Space: Results and Prospects," *Crystallography Reports* 59 (2014): 781, https://link.springer.com/article/10.1134/S1063774514060285; K. W. Benz and P. Dold, "Crystal Growth under Microgravity: Present Results and Future Prospects towards the International Space Station," *Journal of Crystal Growth* 237–39 (2002): part 3, 1638–45.

53. Heinrich M. Jaeger, Sidney R. Nagel, and Robert P. Behringer, "The Physics of Granular Materials," *Physics Today* 49 (1996): 4, 32, doi:10.1063/1.881494.

54. Luis Zea et al., "A Molecular Genetic Basis Explaining Altered Bacterial Behavior in Space," *PLOS One* 11, no. 11 (2016): e0164359, doi:10.1371/journal.pone.0164359.

55. Techshot, "Success: 3D Bioprinter in Space Prints with Human Heart Cells," press release, 7 January 2020, https://techshot.com/success-3d-bioprinter-in-space-prints-with-human-heart-cells/; Michael Molitch-Hou, "Bioprinter Preps for ISS to

3D Print Beating Heart Tissue," engineering.com, 12 July 2018, https://www.engi-neering.com/3DPrinting/3DPrintingArticles/ArticleID/17268/Bioprinter-Preps-for-ISS-to-3D-Print-Beating-Heart-Tissue.aspx.

56. "Biotechnology market size surpassed USD 417 billion in 2018 and is projected to achieve 8.3% CAGR up to 2025." Sumant Ugalmugle and Rupali Swain, "Biotechnology Market Size by Application (Biopharmacy, Bioservices, Bioagriculture, Bioindustries, Bioinformatics), by Technology (Fermentation, Tissue Engineering and Regeneration, PCR Technology, Nanobiotechnology, Chromatography, DNA Sequencing, Cell Based Assay), Industry Analysis Report, Regional Outlook, Application Potential, Competitive Market Share & Forecast, 2019–2025," *Global Market Insights,* Report ID GM1784, November 2018, https://www.gminsights.com/industry-analysis/biotechnology-market.

57. Ashley Strickland, "Why NASA Sent a Superbug to the Space Station," *CNN,* 20 February 2017, https://edition.cnn.com/2017/02/17/health/superbug-mrsa-space-station/.

58. Yves-A. Grondin, "Affordable Habitats Means More Buck Rogers for Less Money, Says Bigelow," Spaceflight.com, 7 February 2014, https://www.nasaspaceflight.com/2014/02/affordable-habitats-more-buck-rogers-less-money-bigelow/.

59. *National Institutes of Health (NIH) Budget—Research for the People*: https://www.nih.gov/about-nih/what-we-do/budget.

60. Ashley Strickland, "NASA-SpaceX Launches Will Boost Science Research on the Space Station," *CNN,* 2 June 2020, https://us.cnn.com/2020/06/02/us/space-station-science-spacex-nasa-scn/index.html.

61. F. D. Gregory, J. J. Rothenberg, et al., "Preparing for the High Frontier: The Role and Training of NASA Astronauts in the Post–Space Shuttle Era," *National Research Council* (2011), 45, https://www.nap.edu/read/13227/chapter/1.

62. "It is the great multiplication of the productions of all the different arts, in consequence of the division of labour, which occasions, in a well-governed society, that universal opulence which extends itself to the lowest ranks of the people." Adam Smith, *The Wealth of Nations,* book 1, chapter 1, p. 22, para. 10, Adam Smith Institute, https://www.adamsmith.org/adam-smith-quotes/.

63. Mike Wall, "1st Private Space Station Will Become an Off-Earth Manufacturing Hub," Space.com, 5 June 2017, https://www.space.com/37079-axiom-commercial-space-station-manufacturing.html.

64. The chemical formula for ZBLAN is ZrF_2-BaF_2-$LaF3$-$AlF3$-NaF. See Edwin C. Ethridge, Dennis S. Tucker, William Kaukler, and Basil Antar, "Mechanisms for the Crystallization of ZBLAN," in *2002 Microgravity Materials Science Conference,* 1 February 2003, 211, NASA Technical Reports Server (NTRS), https://ntrs.nasa.gov/archive/nasa/casi.ntrs.nasa.gov/20030060502.pdf.

65. Ioana Cozmuta and Daniel J. Rasky, "Exotic Optical Fibers and Glasses: Innova-tive Material Processing Opportunities in Earth's Orbit," *New Space* 5, no. 3 (2017): 121.

66. Michel Poulain, Marcel Poulain, and Jacques Lucas, "Verres fluores au tetraflu-orure de zirconium proprietes optiques d'un verre dope au Nd3+," *Materials Research Bulletin* 10, no. 4 (April 1975): 243–46.

67. Debra Werner, "Sparking the Space Economy," Aerospace America, AIAA, January 2020, https://aerospaceamerica.aiaa.org/features/sparking-the-space-economy/.

68. "ZBLAN Continues to Show Promise," NASA Science, 5 February 1998, https://science.nasa.gov/science-news/science-at-nasa/1998/msad05feb98_1.

69. Debra Werner, "FOMS Reports High-Quality ZBLAN Production on ISS," Space News, 7 November 2019, https://spacenews.com/foms-reports-high-quality-zblan-production-on-iss/; Physical Optics Corporation (POC), *Orbital Fiber Optic Production Module: ORFOM,* https://www.poc.com/emerging-technologies/ (accessed 22 June 2020).

70. Gerard K. O'Neill, *The High Frontier: Human Colonies in Space* (New York: William Morrow, 1976).

71. William Yardley, "Peter Glaser, Who Envisioned Space Solar Power, Dies at 90," *New York Times,* 5 June 2014, https://www.nytimes.com/2014/06/06/us/peter-glaser-who-envisioned-space-solar-power-dies-at-90.html; Peter Glaser, "Power from the Sun: Its Future," *Science* 162 (1968): 857, https://science.sciencemag.org/content/162/3856/857.

72. John C. Mankins, *SPS-ALPHA: The First Practical Solar Power Satellite via Arbitrarily Large Phased Array,* section 2.1, 2011–2012 NASA NIAC Phase Project, Artemis Innovation Management Solutions LLC, 5 September 2012, https://www.nasa.gov/sites/default/files/atoms/files/niac_2011_phasei_mankins_spsalpha_tagged.pdf.

73. John Hickman, "The Political Economy of Very Large Space Projects," *Journal of Evolution and Technology* 4 (November 1999), https://jetpress.org/volume4/space.htm.

74. David Szondy, "X-37B Spaceplane Experiment to Test Tech for Beaming Solar Power to Earth," New Atlas, 18 May 2020, https://newatlas.com/space/x-37b-solar-energy-beaming-experiment/; Sandra Erwin, "Navy's Solar Power Satellite Hardware to Be Tested in Orbit," Space News, 18 May 2020, https://spacenews.com/navys-solar-power-satellite-hardware-to-be-tested-in-orbit/.

75. James Conca, "How the U.S. Navy Remains the Masters of Modular Nuclear Reactors," *Forbes,* 23 December 2019, https://www.forbes.com/sites/james-conca/2019/12/23/americas-nuclear-navy-still-the-masters-of-nuclear-power/#36f3460b6bcd.

76. Andrew J. Hawkins, "Electric Flight Is Coming, but the Batteries Aren't Ready—Flying Requires an Incredible Amount of Energy, and Batteries Are Too Heavy," The Verge, 14 August 2018, https://www.theverge.com/2018/8/14/17686706/electric-airplane-flying-car-battery-weight-green-energy-travel.

77. Brian Wang, "Firmamentum, Division of Tethers Unlimited, Gets Contract for Demo of On Orbit Spiderfab Manufacturing Which Will Revolutionize Space Construction," Next Big Future, 25 October 2016, https://www.nextbigfuture.com/2016/10/firmamentum-division-of-tethers.html.

78. Oliver Morton, *The Planet Remade: How Geoengineering Could Change the World* (Princeton, NJ: Princeton University Press, 2015).

79. "Sound of Music Script—Dialogue Transcript," Drew's script-o-rama, http://www.script-o-rama.com/movie_scripts/s/sound-of-music-script-transcript.html (accessed 15 July 2020).

80. Space Adventures: https://spaceadventures.com.

81. Ansari XPRIZE: https://ansari.xprize.org/prizes/ansari; Jess Righthand, "October 4, 2004: SpaceShipOne Wins $10 Million X Prize," smithsonianmag.com, 4 October 2010, https://www.smithsonianmag.com/smithsonian-institution/october-4-2004-spaceshipone-wins-10-million-x-prize-1294605/.

82. Mike Wall, "Ticket Price for Private Spaceflights on Virgin Galactic's Space-ShipTwo Going Up," Space.com, 30 April 2013, https://www.space.com/20886-virgin-galactic-spaceshiptwo-ticket-prices.html.

83. Blue Origin, *Historic Rocket Landing November 23, 2015, West Texas Launch Site,* YouTube, https://www.youtube.com/watch?v=9pillaOxGCo.

84. McDowell, "The Edge of Space."

85. *Gravity,* Internet Movie Script Database (IMSDb), https://www.imsdb.com/scripts/Gravity.html (accessed 7 January 2019).

86. Jeff Foust, "Sierra Nevada Explores Other Uses of Dream Chaser," Space News, 14 January 2020, https://spacenews.com/sierra-nevada-explores-other-uses-of-dream-chaser/.

87. Jeff Foust, "XCOR Aerospace Files for Bankruptcy," Space News, 9 November 2017, https://spacenews.com/xcor-aerospace-files-for-bankruptcy/; Mark Harris, "The Short Life and Death of a Space Tourism Company," *Air & Space Magazine,* December 2017, https://www.airspacemag.com/space/fate-of-the-lynx-180967118/.

88. Douglas Messier, "Bankrupt Spaceflight Company's Space Plane Assets to Help Young Minds Soar," Space.com, 20 April 2018, https://www.space.com/40352-xcor-aerospace-lynx-space-plan-stem-education.html.

89. Kenneth Chang, "There Are 2 Seats Left for This Trip to the International Space Station," *New York Times,* 5 March 2020, updated 17 April 2020, https://www.nytimes.com/2020/03/05/science/axiom-space-station.html.

90. "Living in Outer Space, Space Stations, Chapter 3," Science Clarified, 2020, http://www.scienceclarified.com/scitech/Space-Stations/Living-in-Outer-Space.html.

91. Bonnie Burton, "New ISS Toilet Provides 'Increased Crew Comfort and Performance,' " c|net, 16 June 2020, https://www.msn.com/en-us/news/technology/international-space-station-is-getting-a-toilet-upgrade/ar-BB15xvsx.

92. Tibor S. Balint and Chang Hee Lee, "Pillow Talk: Curating Delight for Astronauts," 69th International Astronautical Congress (IAC), 1–5 October 2018, session E5.3.3,https://www.researchgate.net/publication/328136989_Pillow_Talk-Curating_Delight_for_Astronauts.

93. W. David Compton and Charles D. Benson, "Living and Working in Space," in *A History of Skylab: Living and Working in Space* (1983), NASA History Series, Scientific and Technical Information Office, SP-4208, https://history.nasa.gov/SP-4208/ch7.htm#t4.

94. Take a look at the International Space Station interior photos at Photos House and Design Interior: http://photonshouse.com/international-space-station-interior-photos.html.

95. "To Boldly Brew: Italian Astronaut Makes First Espresso in Space," *Guardian*, 4 May 2015, https://www.theguardian.com/world/2015/may/04/space-italy-coffee-astronaut-espresso-cristoforetti.

96. Mike Wall, "How to Drink Champagne in Space," Space.com, 15 June 2018, https://www.space.com/40900-space-champagne-mumm-bottle-glasses.html.

97. Private communication, 1 May 2019; and from Ed Lu's blog post "Flying" in *Ed's Musings from Space*, 28 July 2003, https://spaceflight.nasa.gov/station/crew/exp7/luletters/lu_letter2.html.

98. Makoto Arai, "Future Prospects and Philosophy of Sports in Space," *69th International Astronautical Congress* (IAC), 1–5 October 2018, session E1.9.5, https://iafastro.directory/iac/paper/id/48580/abstract-pdf/IAC-18.

99. J. K. Rowling, "Quidditch," in *Harry Potter and the Philosopher's Stone* (*Sorcerer's Stone* in the U.S. edition) (London: Bloomsbury, 1997). See also "Quidditch," Harry Potter Wiki, https://harrypotter.fandom.com/wiki/Quidditch; "Quidditch Pitch," Harry Potter Wiki, https://harrypotter.fandom.com/wiki/Quidditch_pitch.

100. Mike Wall, "SpaceX Will Fly a Japanese Billionaire (and Artists, Too!) around the Moon in 2023," Space.com, 18 September 2018, https://www.space.com/41854-spacex-unveils-1st-private-moon-flight-passenger.html.

101. MIT Media Lab Space Exploration Initiative: https://www.media.mit.edu/groups/space-exploration/overview/.

102. Jane Flanagan, "American Woman Killed by Lion Recorded Her Own Grisly Death," *Mail Online*, 2 June 2015, https://www.dailymail.co.uk/news/article-3107036/American-woman-eaten-lion-recorded-grisly-death-Police-examine-camera-tourist-22-took-pictures-beast-approaching-open-car-window-seconds-pounced.html.

103. Out of over 1,800 jumps, there were 97 recorded fatalities from 11 April 1981 to 6 February 2013. See "Fatality List of People Who Dies during BASE Jump," *BASE Jumping*, http://base-jumping.eu/base-jumping-fatality-list/ (accessed 5 September 2020).

104. James A. Vedda, *Becoming Spacefarers: Rescuing America's Space Program* ([Bloomington, IN:] Xlibris, 2012), 103–8.

105. Robert G. Pushkar, "Comet's Tale: A Half Century Ago, the First Jet Airliner Delighted Passengers with Swift, Smooth Flights until a Fatal Structural Flaw Doomed Its Glory," *Smithsonian Magazine*, June 2002, https://www.smithsonianmag.com/history/comets-tale-63573615/.

106. The Boeing 707 had 148 crashes with a total of 2,752 dead. Aviation Safety Network: http://aviation-safety.net/database/type/type-stat.php?type=100.

107. The 282 votes came in as: 14 percent astronauts, 28 percent spaceflight participants, 33 percent CG ballast, 25 percent the 3 percenters. ParabolicArc.com, "What do we call passengers who fly to 80 km, which requires a small percentage of the energy

needed to get to orbit?" Twitter, 12 December 2018, https://twitter.com/spacecom/status/1072895806628225026?s=11.

108. United Nations Office for Outer Space Affairs, *Agreement on the Rescue of Astronauts, the Return of Astronauts and the Return of Objects Launched into Outer Space*, https://www.unoosa.org/oosa/en/ourwork/spacelaw/treaties/introrescueagreement.html (accessed 5 September 2020).

9. Making Space Safe for Capitalism

1. William H. Goetzmann, *Exploration & Empire: The Explorer and the Scientist in the Winning of the American West* (New York: Knopf, 1966), chapter 1, section 5.

2. Marianna Mazzucato, *The Entrepreneurial State: Debunking Public vs. Private Sector Myths* (New York: Public Affairs, 2015).

3. Jeff Foust, "NASA Adjusting its Strategy for LEO Commercialization," Space News, 21 April 2020, https://spacenews.com/nasa-adjusting-its-strategy-for-leo-commercialization/.

4. Alexander MacDonald, *The Long Space Age: The Economic Origins of Space Exploration from Colonial America to the Cold War* (New Haven, CT: Yale University Press, 2017).

5. H.R. Report 1022, George E. Brown, Jr., Near-Earth Object Survey Act, 109th Cong. (2005–6), www.GovTrack.us, https://www.govtrack.us/congress/bills/109/hr1022.

6. "Space is big. You just won't believe how vastly, hugely, mind-bogglingly big it is. I mean, you may think it's a long way down the road to the chemist's, but that's just peanuts to space." Douglas Adams, *The Hitchhiker's Guide to the Galaxy* (BBC Radio, 1978). Numerically, the distance from Earth to Mars at their closest approach is 1,350 times longer than a round-the-world voyage (~40,000 km). Compared to the outer planets, Mars is a near neighbor.

7. John F. Kennedy Moon Speech—Rice Stadium, NASA, 12 September 1962, https://er.jsc.nasa.gov/seh/ricetalk.htm.

8. Karen Cramer, "The Lunar Users Union—An Organization to Grant Land Use Rights on the Moon in Accordance with the Outer Space Treaty," International Institute for Space Law, *Proceedings of the Fortieth Colloquium on the Law of Outer Space* 4, no. 13 (1997): 352–57.

9. Alanna Krolikowski and Martin Elvis, "Making Policy for New Asteroid Activities: In Pursuit of Science, Settlement, Security, or Sales?" *Space Policy* 47 (2019): 7–17.

10. The Space Resources Roundtable, Planetary & Terrestrial Mining Sciences Symposium, ISRU Info: Home of the Space Resources Roundtable, 11–14 June 2019, https://isruinfo.com/public/index.php?page=srr_20_ptmss.

11. Jonathan C. McDowell, "The Low Earth Orbit Satellite Population and Impacts of the SpaceX Starlink Constellation," *Astrophysical Journal Letters*, 892, no. 2 (2020): L36, doi:10.3847/2041-8213/ab8016.

12. Associated Press and Phoebe Weston, "There's a Disco-Ball in Space! Entrepreneur Fires Man-made 'STAR' into Orbit That Will Be the 'Brightest Object in the

Sky': Here's When and Where You Can See It," *Daily Mail On-line*, 25 January 2018, https://www.dailymail.co.uk/sciencetech/article-5308317/Disco-nights-Rocket-Lab-launches-glinting-sphere-orbit.html.

13. Sarah Scoles, "Space Billboards Are Just the Latest Orbital Stunt," Wired, 18 January 2019, https://www.wired.com/story/space-billboards-are-just-the-latest-orbital-stunt/.

14. Aristos Georgiou, "Russian Start-up Plans Massive Space Billboards to Beam Ads to Earth, Dismisses Astronomer Concerns: 'Haters Gonna Hate,' " *Newsweek*, 17 January 2019, https://www.newsweek.com/space-billboard-russian-start-startrocket-orbital-display-advertising-1295422.

15. Inter-Agency Space Debris Coordination Committee: https://www.iadc-home.org.

16. International Telecommunication Union: https://www.itu.int/en/about/Pages/default.aspx.

17. World Intellectual Property Organization, "PCT—The International Patent System," https://www.wipo.int/pct/en/ (accessed 5 September 2020).

18. International Institute of Space Law (IISL): https://iislweb.org/about-the-iisl/introduction/.

19. United Nations Committee on the Peaceful Uses of Outer Space: https://www.unoosa.org/oosa/en/ourwork/copuos/index.html.

20. Francis Lyall and Paul B. Larsen, *Space Law: A Treatise*, 2nd ed. (London: Taylor & Francis, 2017); Tanja Masson-Zwaan and Mahulena Hofmann, *Introduction to Space Law*, 4th ed. (Alphen aan den Rijn: Wolters Kluwer, 2019).

21. Scott Ervin, "Law in a Vacuum: The Common Heritage Doctrine in Outer Space Law," *Boston College International and Comparative Law Review* 7, no. 2 (1984), http://lawdigitalcommons.bc.edu/iclr/vol7/iss2/9.

22. UN Office of Outer Space Affairs, *Treaty on Principles Governing the Activities of States in the Exploration and Use of Outer Space, Including the Moon and Other Celestial Bodies*, 1967, https://www.unoosa.org/oosa/en/ourwork/spacelaw/treaties/introouterspacetreaty.html.

23. "Soviet Robot-Retrieved Moon Rocks Sell for $855,000 at Sotheby's," collect space, November 2018, http://www.collectspace.com/news/news-112918a-sothebys-moon-rock-auction.html.

24. Robert Mccoppin, "NASA Returns Priceless Bag of Moon Dust to Chicago-Area Woman After Lawsuit," *Seattle Times*, 28 February 2017, https://www.seattletimes.com/nation-world/nasa-returns-moon-dust-to-illinois-woman-who-bought-it-at-auction/.

25. Virgiliu Pop, *Who Owns the Moon? Extraterrestrial Aspects of Land and Mineral Resources Ownership* (New York: Springer, 2008).

26. Cody Knipfer, "Revisiting 'Non-interference Zones' in Outer Space," Space Review, 29 January 2018, http://thespacereview.com/article/3418/1.

27. Laura Montgomery, "The 'Non-interference' Provision of Article IX of the Outer Space Treaty and Property Rights," Ground Based Space Matters, 31 March 2017, https://groundbasedspacematters.com/index.php/2017/03/31/the-non-interference-provision-of-article-ix-of-the-outer-space-treaty-and-property-rights/.

28. This is the substance of the law:

§ 51303. Asteroid resource and space resource rights
A United States citizen engaged in commercial recovery of an asteroid resource or a space resource under this chapter shall be entitled to any asteroid resource or space resource obtained, including to possess, own, transport, use, and sell the asteroid resource or space resource obtained in accordance with applicable law, including the international obligations of the United States.
SEC. 403. DISCLAIMER OF EXTRATERRITORIAL SOVEREIGNTY.
It is the sense of Congress that by the enactment of this Act, the United States does not thereby assert sovereignty or sovereign or exclusive rights or jurisdiction over, or the ownership of, any celestial body.

https://congress.gov/congressional-report/109th-congress/house-report/158/1.

29. Andrew Silver, "Luxembourg Passes First EU Space Mining Law. One Can Possess the Spice. Paves Way for Thousands of Sci-Fi Novel Prologues to Come True," Register, 14 July 2017, https://www.theregister.co.uk/2017/07/14/luxembourg_passes_space_mining_law/.

30. L. Barnard, "UAE to Finalise Space Laws Soon," thenational.ae, 7 March 2016, https://www.thenational.ae/business/uae-to-finalise-space-laws-soon-1.219966.

31. Myres S. McDougal, "The Emerging Customary Law of Space," *Conference on the Law of Space and of Satellite Communications*, 1 May 1963, https://digitalcommons.law.yale.edu/cgi/viewcontent.cgi?article=3567&context=fss_papers.

32. The Hague International Space Resources Governance Working Group: https://www.universiteitleiden.nl/en/law/institute-of-public-law/institute-for-air-space-law/the-hague-space-resources-governance-working-group; *Building Blocks for the Development of an International Framework on Space Resource Activities,* 2019, https://www.universiteitleiden.nl/binaries/content/assets/rechtsgeleerdheid/instituut-voor-publiekrecht/lucht--en-ruimterecht/space-resources/bb-thissrwg--cover.pdf.

33. "Planning Pays Off—Chandra Sails through the Leonids Unharmed," Chandra Chronicles, 18 November 2001, https://chandra.harvard.edu/chronicle/0401/leonids_part2.html.

34. Tom Cassauwers, "Is the US Military Spending More on Space Than NASA?" OZY, 24 January 2019, https://www.ozy.com/news-and-politics/is-the-us-military-spending-more-on-space-than-nasa/92148/.

35. The Global Positioning System: https://www.gps.gov/systems/gps/.

36. Immanuel Kant, *Perpetual Peace, a Philosophical Essay* (1795), trans. Mary Campbell Smith (1903), available at Project Gutenberg, https://www.gutenberg.org/files/50922/50922-h/50922-h.htm.

37. Oona A. Hathaway and Scott J. Shapiro, *The Internationalists: How a Radical Plan to Outlaw War Remade the World* (New York: Simon & Schuster, 2017). See review in the *Guardian,* 12 September 2017, https://www.theguardian.com/books/2017/dec/16/the-internationalists-review-plan-outlaw-war.

38. Rare Earth Metals, Mineralprices.com: https://mineralprices.com/rare-earth-metals/.

39. James Vincent, "China Can't Control the Market in Rare Earth Elements Because They Aren't All That Rare," The Verge, 17 April 2018, https://www.theverge.com/2018/4/17/17246444/rare-earth-metals-discovery-japan-china-monopoly; Interview with Kristin Vekasi, "China's Control of Rare Earth Metals," 13 August 2019, National Bureau of Asian Research: https://www.nbr.org/publication/chinas-control-of-rare-earth-metals/.

40. G. Jeffrey Taylor, *A New Moon for the Twenty-First Century,* Planetary Science Research Discoveries, 31 August 2000, http://www.psrd.hawaii.edu/Aug00/new-Moon.html.

41. "It Will Soon Be Possible to Send a Satellite to Repair Another," *Economist,* 24 November 2018, https://www.economist.com/science-and-technology/2018/11/24/it-will-soon-be-possible-to-send-a-satellite-to-repair-another.

42. Cameron Hunter and Bleddyn Bowen, "Donald Trump's Space Force Isn't as New or as Dangerous as It Seems," Conversation, 15 August 2018, https://theconversation.com/donald-trumps-space-force-isnt-as-new-or-as-dangerous-as-it-seems-101401; Sarah Lewin, "Trump Orders Space Force for 'American Dominance,' " Space.com, June 18, 2018.

43. Stephen Kinzer, "Trump's Space Force Is a Silly but Dangerous Idea," *Boston Globe,* 1 September 2018, updated 2 September 2018, https://www.bostonglobe.com/opinion/2018/09/01/trump-space-force-silly-but-dangerous-idea/SPOrX-08QH56ajecVMw9IsI/story.html?p1=Article_Inline_Text_Link. p. K6.

44. Ryan Parry, "Don't Call It Space Force Mr Trump!" *Daily Mail,* 23 January 2019, https://www.dailymail.co.uk/news/article-6619563/Buzz-Aldrin-says-Trump-rename-Space-Force-Space-Guard-aggressive.html.

45. Hugo Grotius, *De iure belli ac pacis (On the Law of War and Peace)* (1625), https://plato.stanford.edu/entries/grotius/#JusWarDoc.

46. "What Is the MILAMOS Project?" In *Manual on International Law Applicable to Military Uses of Outer Space,* McGill Centre for Research in Air and Space Law, https://www.mcgill.ca/milamos/ (accessed 5 September 2020).

47. "Conflict in Outer Space Will Happen: Legal Experts," *University of Adelaide News,* 10 April 2018, https://www.adelaide.edu.au/news/news99182.html.

48. Space Security Index: http://spacesecurityindex.org.

49. Kenneth Chang, "Rocket Lab's Modest Launch Is Giant Leap for Small Rocket Business," *New York Times,* 12 November 2018, B2, https://www.nytimes.com/2018/11/10/science/rocket-lab-launch.html.

50. Erika Ilves, *Make a Dent in the Universe,* TEDxStavanger (2013), YouTube, https://www.youtube.com/watch?v=K89nP7GhWgU.

51. The attribution to Churchill is popular. For example, Ben Smith, "Get It Right! Max's Dad Is Not WSC; WSC Is Misquoted Again," *Atlanta Journal-Constitution,* 29 August 2007, https://www.winstonchurchill.org/publications/finest-hour/finest-hour-136/media-matters/. But this is probably wrong, with a 1938 Budweiser ad being

the likely source. See Quote Investigator: https://quoteinvestigator.com/2013/09/03/success-final/.

52. Bruce Watson, "Mr. Feynman Goes to Washington," *Attic* (1986), https://www.theattic.space/home-page-blogs/Feynman.

53. Tariq Malik, "Elon Musk Explains Why SpaceX's Falcon Heavy Core Booster Crashed," Space.com, 14 February 2018, https://www.space.com/39690-elon-musk-explains-falcon-heavy-core-booster-crash.html.

54. Jeff Foust, "10 Years and Counting—A Decade After X Prize Victory, Suborbital Service Still on the Cusp," Space News, 13 October 2014, https://spacenews.com/42171110-years-and-counting-a-decade-after-x-prize-victory-suborbital-service-still/.

55. Elizabeth Howell, "Facts about SpaceX's Falcon Heavy Rocket," Space.com, 22 February 2018, https://www.space.com/39779-falcon-heavy-facts.html.

56. *Platinum Prices—Interactive Historical Chart*, Macrotrends, https://www.macrotrends.net/2540/platinum-prices-historical-chart-data.

57. Carol Dahl, Ben Gilbert, and Ian Lange, "Mineral Scarcity on Earth: Are Asteroids the Answer," *Mineral Economics* 33 (2020): 29–41.

58. "The estimated global platinum production in 2019 was 180 metric tons." M. Garside, "Global Platinum Mine Production by Country, 2015–2019," Statista.com, 13 February 2020, https://www.statista.com/statistics/273645/global-mine-production-of-platinum/.

10. In the Long Run

1. Erika Ilves, *Make a Dent in the Universe,* TEDxStavanger (2013), YouTube, https://www.youtube.com/watch?v=K89nP7GhWgU; Harry L. Shipman, *Humans in Space: 21st Century Frontiers* (New York: Springer, 1989).

2. Melissa Dell, "The Persistent Effects of Peru's Mining Mita," *Econometrica* 78, no. 6 (2010): 1863–1903, https://scholar.harvard.edu/files/dell/files/ecta8121_0.pdf.

3. "The Doctrine of First Effective Settlement": "Whenever an empty territory undergoes settlement, or an earlier population is dislodged by invaders, the specific characteristics of the first group able to effect a viable, self-perpetuating society are of crucial significance for the later social and cultural geography of the area, no matter how tiny the initial band of settlers may have been. Thus, in terms of lasting impact, the activities of a few hundred, or even a few score, initial colonizers can mean much more for the cultural geography of a place than the contributions of tens of thousands of new immigrants a few generations later." Wilbur Zelinsky, *The Cultural Geography of the United States* (Englewood Cliffs, NJ: Prentice-Hall, 1973), 13–14.

4. Gilbert White, *A Natural History of Selborne* (1789). See https://naturalhistoryofselborne.com (accessed 21 September 2020). Mr. Bayes and Mr. Price, "An Essay towards Solving a Problem in the Doctrine of Chances. By the Late Rev. Mr. Bayes, F. R. S., Communicated by Mr. Price, in a Letter to John Canton, A. M. F. R. S," *Philosophical Transactions of the Royal Society of London* 53 (1763): 370–418, doi:10.1098/rstl.1763.0053.

5. Professor Pangloss was the teacher of Candide, the title character in Voltaire's 1759 novella, which ridicules the professor's optimistic view that "All's for the best in the best of all possible worlds," based on the philosopher Gottfried Leibnitz's work. Leibnitz was also the co-inventor of calculus.

6. Carl Sagan, "Blues for a Red Planet," in *Cosmos* (New York: Penguin Random House, 1980), https://www.liquisearch.com/carl_sagan/personal_life_and_beliefs.

7. Robert Zubrin, president of the Mars Society, is a notable case: "I think Sagan's statement is basically political correctness gone berserk. It's completely wrong. Ethics needs to be based on what's best for humanity, not what's best for bacteria." Jason Koebler, "If Curiosity Finds Life on Mars, Then What?" *U.S. News and World Report*, 17 August 2012, https://www.usnews.com/news/articles/2012/08/17/if-curiosity-finds-life-on-mars-then-what.

8. Leonard David, "Jeff Bezos' Vision: 'A Trillion Humans in the Solar System,' " Space.com, 21 July 2017, https://www.space.com/37572-jeff-bezos-trillion-people-solar-system.html.

9. B. Gladman et al., "The Structure of the Kuiper Belt: Size Distribution and Radial Extent," *Astronomical Journal* 122 (2001): 1051–66, https://www-n.oca.eu/morby/papers/PencilB.pdf.

10. Paul Krugman, "The Theory of Interstellar Trade," *Economic Inquiry* 48 (2010): 1119–23, http://www.standupeconomist.com/pdf/misc/interstellar.pdf.

11. Richard P. Feynman, from a transcript of "Seeking New Laws," the seventh Messenger Lecture, Cornell University (1964), published in *The Character of Physical Law* (1965; repr., Cambridge, MA: MIT Press, 1967), 172.

12. K. Batygin and M. E. Brown, "Evidence for a Distant Giant Planet in the Outer Solar System," *Astronomical Journal* 151 (2016): 22, https://iopscience.iop.org/article/10.3847/0004-6256/151/2/22. Mike Brown discovered Eris, a body in the Kuiper Belt originally thought to be bigger than Pluto, setting off the crisis of whether or not Pluto was a true planet. He is proud of this; his Twitter handle is @plutokiller. He tells the story in his book *How I Killed Pluto and Why It Had It Coming* (New York: Spiegel & Grau, 2010).

13. Jakub Scholtz and James Unwin, "What if Planet 9 Is a Primordial Black Hole?" *Physical Review Letters* 125, no. 5 (2020).

14. Miguel Alcubierre, "The Warp Drive: Hyper-fast Travel within General Relativity," *Classical and Quantum Gravity* 11, no. 5 (1994): L73–L77, doi:10.1088/0264-9381/11/5/001.

15. Phil Dooley, "Could 'Negative Mass' Unify Dark Matter, Dark Energy?" *Cosmos Magazine*, December 2018, https://cosmosmagazine.com/space/could-negative-mass-unify-dark-matter-dark-energy; J. S. Farnes, "A Unifying Theory of Dark Energy and Dark Matter: Negative Masses and Matter Creation within a Modified ΛCDM Framework," *Astronomy & Astrophysics* 620 (2018): L11. Farnes's theory does also need the continuous creation of negative mass particles to get cosmic acceleration. Some would argue that is invoking the tooth fairy too many times.

16. E. Witten, "Cosmic Separation of Phases," *Physical Review D* 30 (1984): 272–85.

17. Feryal Ozel et al., "The Dense Matter Equation of State from Neutron Star Radius and Mass Measurements," *Astrophysical Journal* 820 (2016): 28.

18. Charles Alcock, "Engineering with Quark Matter," *Nature* 337, no. 6206 (1989): 405, doi:10.1038/337405a0. The Three Gorges Dam has a maximum power output of 22.5 gigawatts. "Three Gorges Dam Hydro Electric Power Plant, China," *Power Technology,* https://www.power-technology.com/projects/gorges/ (accessed 5 September 2020).

19. J. E. Horvath, "The Search for Primordial Quark Nuggets among Near Earth Asteroids," *Astrophysics and Space Science* 315, nos. 1–4 (2008): 361–64.

20. Charles Alcock, Edward Farhi, and Angela Olinto, "Strange Stars," *Astrophysical Journal* 310 (1986): 261.

21. American Institute of Physics, "A View from the White House: Marburger on S&T Funding Priorities," 21 February 2003, no. 23, https://www.aip.org/fyi/2002/view-white-house-marburger-st-funding-priorities.

22. As shown on their mission patch: http://lroc.sese.asu.edu/about/patch.

23. Steve Squyres, *Roving Mars: Spirit, Opportunity, and the Exploration of the Red Planet* (New York: Hyperion, 2005).

24. Claudia Dreifus, "A Conversation with Sir Martin Rees; Tracing Evolution of Cosmos from Its Simplest Elements," *New York Times,* 28 April 1998, https://www.nytimes.com/1998/04/28/science/conversation-with-sir-martin-rees-tracing-evolution-cosmos-its-simplest-elements.html.

25. "Proposed Missions—Terrestrial Planet Finder," *JPL,* 19 June 2003, https://www.jpl.nasa.gov/spaceimages/details.php?id=PIA04499.

26. Emerging Technology from the arXiv, "A Space Mission to the Gravitational Focus of the Sun," *MIT Technology Review,* 26 April 2016, https://www.technology-review.com/2016/04/26/8417/a-space-mission-to-the-gravitational-focus-of-the-sun/; Geoffrey A. Landis, "Mission to the Gravitational Focus of the Sun: A Critical Analysis," *AIAA Science and Technology Forum and Exposition* (2017), arXiv:1604.06351.

27. *Breakthrough Starshot:* https://breakthroughinitiatives.org/initiative/3.

28. *Enduring Quests, Daring Visions—NASA Astrophysics in the Next Three Decades,* National Academies, 2013, https://science.nasa.gov/science-committee/subcommittees/nac-astrophysics-subcommittee/astrophysics-roadmap.

29. Center for High Angular Resolution Astronomy, Georgia State University: http://www.chara.gsu.edu/public/tour-overview; Rapid Rotators: http://www.chara.gsu.edu/science-highlights/rapid-rotators.

30. GRAVITY Collaboration, "Spatially Resolved Rotation of the Broad-Line Region of a Quasar at Sub-parsec Scale," *Nature* 563 (2018): 657–60. Based on the concept by Martin Elvis and Margarita Karovska, "Quasar Parallax: A Method for Determining Direct Geometrical Distances to Quasars," *Astrophysical Journal* 581 (2002): L67–L70.

31. Event Horizon Telescope: https://eventhorizontelescope.org/about; The Event Horizon Telescope Collaboration, "First M87 Event Horizon Telescope Results, I: The Shadow of the Supermassive Black Hole," *Astrophysical Journal Letters* 875, no. 1 (2019).

32. William Shakespeare, *King Lear,* act 2, scene 4.

Illustration Credits

Figure 1a, Eros: NASA
Figure 1b, Bennu: NASA
Figure 1c, Kleopatra: NASA/JPL
Figure 1d, Ceres: NASA/JPL
Figure 2, Planetary systems against a gaseous nebula: NASA, Space Telescope Science Institute
Figure 3, Space dust: NASA/JPL
Figure 4, World map of the 190 confirmed impact craters: Data from the Earth Impact Database, Planetary and Space Science Centre, University of New Brunswick, Canada; redrawn by Bill Nelson
Figure 5, Widmanstätten pattern in the Toluca nickel-iron meteorite: H. Raab, Wikipedia user Vesta, CC BY-SA 3.0
Figure 7, Map of hard-rock mines in Colorado: Data from the Colorado Division of Reclamation, Mining and Safety; redrawn by Bill Nelson
Figure 8, Cereal dispenser: Martin Elvis
Figure 9, ZBLAN: NASA/Tucker

Index

Note: Page numbers in italics indicate illustrations.

capitalism in space; manufacturing in
space
space law. *See* regulation (space law)
SpacePharma, 184
"Space Resource Exploration and Utili-
zation Act of 2015," 220
Space Security Index (SSI), 225
SpaceShipOne, 194
SpaceShipTwo, 194–195
Space Shuttles, 161, 226–227
Space Station Alpha, 177, 181
space stations. *See* International Space
Station (ISS)
Spacewatch, 117
SpaceX: *Dragon*, 167–168, 172; *Dragon
2*, 173; *Falcon 9* rockets, 166–170; *Fal-
con Heavy*, 169, 227, 228; *Falcon
Super Heavy*, 170; ISS cargo resupply
services, 172; new rockets and refuel-
ing, 137; passengers, 174, 196; *Star-
ship*, 105, 170, 200, 264n16;
Super-Heavy/Starship, 81
spectroscopy, 18–20, 96, 97, 128, 129
spies and spying, 215–216, 222–223
Spitzer telescope, 4
Springmann, Alessondra, 82
spy satellites, 222
Squyres, Mike, 241
Stark, Philippe, 198
Starliner, 167, 168, 173
Starship, 81, 170
Star Trek (*Enterprise*), 1, 5, 36, 232, 234,
237
StartRocket, 212
Steinhardt, Paul, 89
stock market crash and asteroid mining,
228–229
stony asteroids, 11, 13–15, 128. *See also*
rubble-pile asteroids
Strange, Nathan, 91–92
Stross, Charles, 49
Stuart, Joseph Scott, 128
Stürmer, Martin, 228

Suffredini, Mike, 177, 178
Sun Tzu, 109, 256n1
Super-Heavy/Starship, 81
supernovae, 57
synthetic tracking, 138

Tagish Lake meteorite, 87–88, 103
tailings from space mining, 90–92, 156
tama-hagane steel, 151–152
Tantardini, Marco, 159
Taylor, Anthony, 132
Techshot, 183
telescopes: expense of, 4, 227–228; industrial use of, 138–139; interferometry, 243–244; measuring asteroids,
114–115; NASA successful astronomy
missions, 4, 11, 169–170; and prospecting, 111–112, 146; and search for
near-Earth asteroids, 97, 110–112
Tethers Unlimited, 192
tetrataenite, 85–86
Thales-Alenia modules, 178
thermal inertia, 23, 24
thermal infrared, 22
Thomas-Keprta, Kathie, 45–46
Titov, Gherman, 197
Titus-Bode law, 15, 16
tomography, 98
Tonry, John, 135
tourism, 193–203; motion sickness, 197;
orbital passenger flights, 195–197;
safety issues, 200–202; sports and activities, 199–200; suborbital flights,
194–195; and use of term astronaut,
202
TransAstra, 154, 190–191
Treaty on Principles Governing the …
Use of Outer Space. *See* Outer Space
Treaty (1967)
Trojan asteroids, 38, 103
Trump, Donald, 225
Tsiolkovsky, Konstantin, 131
Tucker, Dennis, 187–188